Advanced Sciences and Technologies for Security Applications

The series Advanced Sciences and Technologies for Security Applications comprises interdisciplinary research covering the theory, foundations and domain-specific topics pertaining to security. Publications within the series are peer-reviewed monographs and edited works in the areas of:

- biological and chemical threat recognition and detection (e.g., biosensors, aerosols, forensics)
- crisis and disaster management
- terrorism
- cyber security and secure information systems (e.g., encryption, optical and photonic systems)
- traditional and non-traditional security
- energy, food and resource security
- economic security and securitization (including associated infrastructures)
- transnational crime
- human security and health security
- social, political and psychological aspects of security
- recognition and identification (e.g., optical imaging, biometrics, authentication and verification)
- smart surveillance systems
- applications of theoretical frameworks and methodologies (e.g., grounded theory, complexity, network sciences, modelling and simulation)

Together, the high-quality contributions to this series provide a cross-disciplinary overview of forefront research endeavours aiming to make the world a safer place.

The editors encourage prospective authors to correspond with them in advance of submitting a manuscript. Submission of manuscripts should be made to the Editor-in-Chief or one of the Editors.

Richard Jiang · Ahmed Bouridane ·
Chang-Tsun Li · Danny Crookes · Said Boussakta ·
Feng Hao · Eran A. Edirisinghe
Editors

Big Data Privacy
and Security in Smart Cities

Springer

Editors
Richard Jiang
School of Computing and Communications
Lancaster University
Lancaster, UK

Chang-Tsun Li 🆔
School of Information Technology Deakin
University Deakin
Geelong, VIC, Australia

Said Boussakta
School of Engineering
Newcastle University
Newcastle upon Tyne, UK

Eran A. Edirisinghe
Keele University
Staffordshire, UK

Ahmed Bouridane
Centre for Data Analytics
and Cybersecurity
University of Sharjah
Sharjah, United Arab Emirates

Danny Crookes
Queen's University Belfast
Belfast, UK

Feng Hao
Department of Computer Science
University of Warwick
Warwick, UK

ISSN 1613-5113   ISSN 2363-9466   (electronic)
Advanced Sciences and Technologies for Security Applications
ISBN 978-3-031-04426-7   ISBN 978-3-031-04424-3   (eBook)
https://doi.org/10.1007/978-3-031-04424-3

This Springer imprint is published by the registered company Springer Nature Switzerland AG
The registered company address is: Gewerbestrasse 11, 6330 Cham, Switzerland

Contents

Smart Cities: A Survey of Tech-Induced Privacy Concerns

Edgard Musafiri Mimo and Troy McDaniel

Abstract Internet of Things (IoT) has become a prominent part of the technologies leveraged by smart cities. IoT enables the collection of data and the processing of information to provide a better value to the city and its citizens. The use of IoT devices in smart cities has enabled many applications that generate security issues despite their provided benefits. These security issues in return generate more citizens' or users' privacy issues as the data and information flow are compromised. There are many privacy related concerns in smart cities that are generated from the security issues pertaining to IoT, Big Data, and ICT enabled tech applications, which require a thorough understanding to better build resilient privacy aware smart cities. This paper provides a comprehensive, characterized, and wide-ranging synopsis of the research on privacy issues springing from IoT, Big Data, and ICT enabled tech applications and their associated security flaws and presents solutions as they pertain to the relevant citizen value driven privacy framework quadrant (Musafiri Mimo and McDaniel in 3D privacy framework: the citizen value driven privacy framework [1]). The characterization is based on the application's most relevant privacy framework quadrant, the applicable security flaws, and the present solutions that potentially lessen the associated privacy concerns among citizens in smart cities.

Keywords Smart cities · Privacy · Framework · Security · Surveillance · IoT · Big Data · ICT · Citizen centered · ITS · Smart energy · Smart governance · Smart health

E. M. Mimo (✉) · T. McDaniel
Arizona State University, Tempe, USA
e-mail: emusafir@asu.edu

T. McDaniel
e-mail: troy.mcdaniel@asu.edu

© Springer Nature Switzerland AG 2022   1
R. Jiang et al. (eds.), *Big Data Privacy and Security in Smart Cities*,
Advanced Sciences and Technologies for Security Applications,
https://doi.org/10.1007/978-3-031-04424-3_1

1   Introduction

With the current growth of the urban population around the world, there is a defi-
nite need for the major cities to transition from just being cities into becoming
relevant smart cities. This is important because of the challenges that arise due to
the overcrowding of cities caused by the migration of citizens from rural to urban
places where most of the provided services tend to exist. The provision of services
and opportunities in smart cities require automation, optimization, and efficiency to
better manage the operations of smart cities with the needed speed and reliability to
satisfy the need of the citizens. These provisions are enabled even more as advances
and progress from IoT, Big Data, and ICT frameworks take place with the technological transformation of the smart cities to meet the present needs and the future
necessities of the cities to ensure citizens thrive.

It is important to consider how technologies and systems that are deployed to
interact with citizens are drastically transforming the way smart cities are shaped
as citizens find numerous accommodating ways to accept and adopt the technologies and the systems. The internet of things (IoT), Big Data, and the information
communication technology (ICT) frameworks form the basis on which smart cities
are built today, and as such, they enable the possibility of having efficient, optimal
and practical smart cities that solve citizens' problems in terms of demand, supply,
and management of their needs and services. However, the practical smart cities that
are built simply based on IoT, Big Data and ICT frameworks have manifested some
lack of effectiveness not because of the problems and issues they are or are not able
to solve, but because of the problems and issues they have generated for citizens
themselves.

The generated problems pertaining to citizens' security and privacy tend to be of
more value to the citizens than the problems that the frameworks intended to solve
in the first place. The generated issues are namely privacy, security and ownership of
the technologies, systems, and their data; all of which are the result of the IoT, Big
Data and ICT frameworks. The issue of effectiveness points and focuses more on the
issues that affect the citizens and their associated long-term effects on the citizens.
The issue of privacy and security is important to all citizens, and as such they should be
considered and addressed not when there are IoT enabled tech applications' problems
generated, but rather during the integration process where these three frameworks
meet and converge to enable smart cities' capabilities.

The advancement and availability of various IoT devices and systems have enabled
the collection of a plethora of types of data in high volume that has enabled more
progress in Big Data as more value becomes available with proper analysis of the
data. The processing of information that ICT systems enable to provide a better value
to the city and its citizens remains one of the two biggest assets of smart cities and
the other is the citizens themselves. The presence and deployment of IoT devices
in smart cities have enabled many applications; many of which are generating a
lot of security issues despite the benefits. These security issues in return generate
more citizens' or users' privacy issues and concerns as the data and the information

flow in the smart cities' ecosystem are compromised. There are many enabled tech application security issues that generate privacy related concerns in smart cities that require in-depth consideration to better build resilient privacy aware smart cities.

Consequently, it is paramount to assess, understand, and reconsider the security and the privacy concerns that are generated by IoT, Big Data and ICT frameworks pertaining to the applications and systems they enable to make smart cities possible. Smart cities must be enabled in a way that preserves the security and privacy of the citizens as recommended by the 3D citizen value driven framework [1] that suggests building citizen privacy aware smart cities for their effective enablement. The objective of this paper is to provide an overview of the various privacy related security issues and some associated present and future solutions that spring forth in literature through the engagement of IoT, Big Data and ICT frameworks in the implementation of smart city technologies as they pertain to the most relevant citizen value driven privacy framework quadrant [1]. The characterization of several smart cities' applications, systems and technologies springing from IoT, Big Data and ICT enabled services and applications in smart cities that present some security induced privacy concerns are discussed to showcase the relevance of the citizen value driven privacy framework quadrant and the available means to lessen the associated privacy related security concerns.

2 Related Work

2.1 Internet of Things (IoT)

IoT remains one of the main enablers of smart cities because of the possibilities of providing avenues to collect the necessary data or information in various ways and forms to enable good and informed decision making in the overall smart cities' transformation. It is very difficult to completely define IoT and its impacts in enabling smart cities, yet its presence is seen and felt almost everywhere in all the sectors of smart cities. Thus, IoT constitutes the core of the smart city's implementation since without its advancements and deployments over the years, many smart cities' initiatives would be almost impossible today. The objective of IoT is to provide ways to collect all types of data in various ways both for structure and unstructured data. The various ways of data collection form a network enabled by the IoT deployment of devices to ensure the smart city possesses the right instrumentation to enable the needed interconnectedness that powers the intelligence of smart cities [2].

This network is built with instrumentation that includes various sensors, actuators, electronics, networks, firmware, middleware, and software. The instrumentation allows for various objects or things like computers, smart phones, wearable devices, homes, buildings, structures, vehicles, and energy systems to become the collectors of data in smart cities, and as such there are multitude of data types and data sources that can be collected and enabled respectively. The services the IoT devices enable are almost everywhere nowadays, and they facilitate the explosion of possibilities pertaining to what can be enabled in cities to truly make them smart in how they operate and provide to the needs and wants of their citizens. The major advantages in having several types of IoT devices are due to affordability, availability, and ease of deployment to enable the collection of data and physically interact harmlessly with the citizens.

It is evident that the security issues surrounding IoT devices are in two folds namely the physical and the technological layer. The physical layer involves the security issues associated with the physical IoT devices and their locations as point of contact and data collection. The technological layer involves the security issues pertaining to the access, control and permission to the collected data or the operation of the IoT devices. The two layers of security give rise to many securities-related worries in smart cities that must be addressed and solutioned long before citizens start to copiously trust and give into the smart cities privacy aware notion since it will eliminate the chances of security induced privacy linked doubts amid citizens [3].

The fact that there are not many robust control measures and approved standards among all smart cities regarding the specific data types that can be captured and recorded by the various IoT devices, for example with audio recorders, deployed in smart cities poses privacy concerns as many citizens may not consent to be recorded or documented in certain environments [3]. It is obvious in image and video camera film collections that are documented, collected, and recorded in smart cities; many times, citizens are not even conscious that they are being recorded on camera, so this is another path of possible security encouraged privacy anxiety that smart cities must address and properly deal with in the effort of creating privacy aware smart cities [3].

Many applications, systems and technologies are being enabled today in smart cities with the accessibility of IoT devices that generate privacy concerns like motion sensing, face recognition, sound detection, biometrics capture and so on since there are security issues pertaining to points of data collection, where there are possibilities to collect data from various people at once even when only a specific individual examination is required [3]. Therefore, the technological security layer presents a bigger privacy danger should a hacker gain access and control of the IoT device at the point of data collection, then citizens' information may be at risk.

Many resolutions have been recommended to help lessen the security and privacy burdens generated by IoT devices in analyzing the collected data information near the point of collection, but still these resolutions do not address the essence of the problems that lie in the criterion of the enabled technologies and the deployment of IoT devices in smart cities in the first place. The use of IoT devices in smart cities remains one of the utmost resources and enabler of opportunities in smart cities, and the future smart cities' resolutions remain severely trusting on the evolution and disposition of abundant IoT devices [3].

2.2 Big Data

Big Data encompasses a way of handling and managing the large and complex amount of data of various types that are collected through the myriad of deployed IoT devices and other avenues in smart cities. Big Data comprises all the different kinds of data both structured and unstructured that require storage, analysis, use and disposition. It is evident that having a massive amount of data in processing for applications and storing for trends and patterns give rise to many issues and concerns especially when personal sensitive data is involved. As such, there are many security and privacy issues that arise in collecting data, processing data, and storing data in smart cities that must be addressed to enable better privacy aware smart cities. There are more than 2.5 quintillion bytes per day of data [2] being collected and stored in one form or another as part of Big Data, and many of the collected data carries personal and sensitive information about citizens who have not consented to having their data collected and stored. The unconsented collection of data leads to privacy concerns when the data or systems housing them are compromised [4].

The need to secure data and the devices that process, store, and dispose of it remains crucial to citizens especially when the information involved is deemed personal. Whenever there is a risk of security being compromised with a potential chance of personal data being tampered with, privacy concerns arise from the citizens that is induced by the security deficiency or breach. In this regard, these security induced privacy concerns would not have been experienced by citizens if Big Data were not collected, processed, and stored in the first place. The transmission and transfer of data from systems to systems can be interrupted or intercepted resulting in data being stolen and exposed by malicious intent. The processing of data can be compromised if the systems are not secured during the analysis or when data is being analyzed with third party software to facilitate enhanced decision making.

The data can also be compromised when it is stored in systems or portable devices that can be duplicated and maliciously accessed to acquire the sensitive data. In view of all these potential factors, there are many security induced privacy concerns that must be addressed as it pertains to Big Data in privacy aware smart cities. The full potential of Big Data is yet to be unfolded to enable even greater value in smart cities with the use of advanced analytics and algorithms that facilitate the retrieval of meaningful information, but it must be achieved with thorough privacy and security considerations that pertain to citizens' information [2, 5]. Big Data include data such as atmospheric data, call detail records, genomic data, e-commerce data, Internet search indexing, medical records, military surveillance, photography archives, RFID data, sensor network data, social network data, video archives, and web logs [2, 5]. From these examples, there is a multitude of security induced privacy concerns that arise because of the sensitivity of this data if it is jeopardized in the process.

2.3   Information and Communication Technology (ICT)

ICT is another pillar that enables many smart cities solutions and developments to ensure processes and systems become autonomous, efficient, and optimal [2]. ICT facilitates the interconnect between Big Data and IoT in the overall smart city infrastructure to enable potential solutions that enhance citizens' quality of life and wellbeing. It is paramount to consider ICT as important as both IoT and Big Data because it is the means of translating and materializing the sound decision made through the interpretation of Big Data that is collected via IoT devices to impact and enhance citizens' life in smart cities. ICT like some IoT devices is the framework that truly interacts with the citizens in ways of implementing the changes that does not harm citizens and implementing changes that citizens can relate to. Almost all the systems and technologies in smart cities are enabled through ICT, which offers a way for citizens to adopt and interact with them.

In this regard, one of the objectives of ICT is to enable citizens' interaction and adoption of systems and technologies as well as facilitate the feedback of citizens back to the system integrators and designers to ensure better ICT infrastructures are developed and deployed to realize smart city solutions. ICT plays a big role in smart cities' sectors through its presence in enabling technologies, and as such there are many avenues for security risks as different sets of systems interact, feed into each other, and share data with different levels of security. The interdependency of systems and protocols within the ICT network poses security problems as data transfer is done in many different forms to facilitate the materialization of the decision making that enables the efficiency and effectiveness of systems and technologies in smart cities. These security issues can lead to many other security induced privacy concerns among citizens especially when there are breaches in communication or unauthorized transfer of personal information.

Thus, it is evident that as more and more interconnected technologies are developed and used in ICT, there are security risks and issues that need to be addressed in view of citizens' privacy concerns that may arise because of compromission that can intentionally or unintentionally occur. Privacy issues pertaining to the security issues caused by ICT network breaches are countless in smart cities including prevalence in elderly's health concerns, public place facial recognition and alteration of mobility systems [2, 5]. The potential use and presence of ICT in smart city space is without a doubt the core enabler of smart city networks and the main layer that interacts with citizens either in receiving information from citizens or performing and completing a service for citizens.

3   Smart Cities Technologies

There are several smart city technologies that can be considered in identifying security induced privacy concerns in all aspect of life in smart cities, where more work is needed to better address these issues if the privacy aware smart cities' status is to become a reality. It is important to ensure that the technologies and the systems powering smart cities preserve the privacy associated with personal data in all interactions, communications, and transfer of data in executing or providing services. Thus, this paper discusses four different technologies enabling smart cities today together with some of their associated systems enabled by IoT, Big Data and ICT frameworks where many security issues and privacy concerns are experienced. These four technologies include smart mobility and transportation, smart energy, smart health, and smart governance.

3.1 Smart Mobility and Transportation Technology

Smart mobility and transportation technology in smart cities is evident and plays a big role in making cities smarter. It is almost impossible for a city to be considered smart if there are no initiatives in addressing the mobility and transportation issues and needs of its citizens because it is one of the main drivers that create problems and needs in many cities in the first place. The migration of people from rural to urban areas are mainly eased through the development of avenues that facilitate the mobility and transportation of people. As the number of citizens exponentially increases in smart cities over the years, there arises a need to heavily invest in improving the mobility and transportation means of cities to facilitate the movement of people within and in and out cities.

Intrinsically, smart cities are investing heavily in systems that enable more robust and resilient mobility and transportation infrastructures that address and provide optimal solutions to transportation dilemmas such as the intelligent transportation system known as ITS, which likewise encompasses several other sub-systems all of which need to be protected and preserve the privacy of personal data flowing through them to effectively enable the convenience that the ITS provides. ITS systems enable smart mobility and transportation technologies in leveraging and integrating sensing opportunities, analysis potential, control capabilities, and means of communication technologies enabled by IoT, Big Data and ICT framework to enhance the mobility, comfort, efficiency, and safety of the ITS systems for the benefit of citizens [4, 5].

Several components come into play in ITS systems such as smart vehicles, public transportation, many IoT devices such as smart phones and other mobile devices, and several controllers such as traffic lights, digital road signs and spot sensor controllers play their share part in enabling what is possible in the transportation and mobility's space in smart cities. ITS components are enabled by several technologies empowered by the integration of IoT, Big Data and ICT frameworks such as sensing technologies that facilitate awareness, quick informed decision making and fast response; computation technologies such as cryptographic computation, fog computing and cloud computing to facilitate the flexibility, availability, reliability and quality of the ITS services; analytic technologies that provide safety and priority based system wide decision making to optimize the ITS; vehicle-to-everything (V2X) technology that facilitates both interactions of vehicle to infrastructure and vehicle to everything through VANETs and RFID; and communication networks technologies to simplify the overall communication protocol within the ITS.

Citizens in smart cities play a big role in using the ITS and advancing its services while interacting and communicating with it using various kinds of IoT devices that they own or that are provided to them through kiosks and machines throughout smart cities. The security issues are therefore plenty in ITS that must be well addressed and controlled to not aggravate privacy concerns among citizens. The heavy use of IoT devices in ITS brings many security issues associated with IoT devices in the ITS space as basic security and privacy standards are not instigated in the IoT devices despite their usefulness in enabling the services in ITS [2, 6]. The sensing and communication network technologies in ITS are susceptible to security vulnerabilities that cannot be ignored or overlooked especially when there is a potential to impact personal data that might be interrupted through communication protocols.

Customers sharing their locations, and vehicles sharing their information in the overall ITS infrastructure that can potentially be intercepted and redirected due to deficiencies in security, require a re-evaluation as to what matters the most in smart cities as it pertains to citizens' information while using ITS services. Several solutions are proposed through research for the different ITS security issues when it comes to smart vehicle security concerning the controller area network (CAN) as implementing methods of encryption and authentication is paramount and should be used [2, 7] to maintain the security and privacy in many application layers of communication [4]. In the case of ECU security, adopting eavesdropping [8] and spoofed packets prevention solutions, and physically securing the vehicle [9], so that there is no room for unauthorize access to the vehicle on-board diagnostics (OBD-II) port to access the ECUs and CAN bus, is paramount.

The same security issues can be experienced with VANETs as the same wireless network technology challenges and attacks pertaining to denial of service, Sybil [10] and replay attacks exist. The privacy solutions in ITS involves using pseudonyms to provide the services while keeping the user fictitious and even anonymous in preserving the privacy of users while ensuring that there are non-repudiation mechanisms that negatively affect the provided service [11–14]. The 3D Privacy Framework recommends to properly assess applications, systems, and processes in smart cities as they pertain to smart mobility and transportation technology to regulate privacy vulnerabilities as to what data is being collected and how to preserve the data to ensure that privacy concerns are lessen where possible.

The 3D framework classifies smart mobility and transportation as it pertains to the city in quadrant 1, and as such it is a high privacy risk space that necessitates adoption of regulation in data collection and options of addressing security issues [1]. Table 1 shows a summary of some prevalent security induced privacy issues together with examples in literature pertaining to smart mobility and transportation technology in smart cities that would have been prevented should IoT, Big Data and ICT frameworks enabling them were meticulously assessed and regulated to prevent excessive collection of personal identifiable data prior to being deployed in smart cities.

Table 1 Security induced privacy concerns pertaining to smart mobility and transportation technology in smart cities

Security issues	Induced privacy concerns	Threat examples	Possible resolution	References
Active secure IoT sensing attack	Authentication, identification	Spoofing in sensing systems	Pseudonym, attribute-based credentials, message authentication codes	[15, 15–18, 65]
Active secure ICT computing attack	Authentication, identification	Race condition/timing attacks	Public-key cryptography, symmetric and asymmetric key, cryptography	[10, 15, 19, 20]
Active and/or passive secure ICT communication network issues	Identification, authentication, confidentiality	Sybil attacks in communication, man in the middle attacks, eavesdropping	Attribute-based credentials, location cloaking, homomorphic encryption, challenge-response protocol, steganography	[10, 15, 21–23, 64]
Active secure Big data machine learning and artificial intelligence issues	Confidentiality, authentication	Model Identification attacks Data intoxication attacks	Signature-based authentication	[16, 18]
Active secure Big Data analytics	Authentication	Data intoxication attacks	Digital signatures, challenge-response protocol	[16, 10, 24]
Passive secure controllers issues	Confidentiality	Parameter/dynamic inference attacks	Signature-based authentication, message authentication codes	[16, 10, 25, 26, 18]

3.2 Smart Energy Technology

Smart energy technology in smart cities is the core that drives almost all other smart technologies in terms of ensuring efficiency and optimization in smart cities. The growing number of citizens in smart cities necessitates a careful consideration of energy consumption and energy distribution to optimally meet the needs of citizens as well as a careful consideration of energy generation avenues to help in the long term as the trend of demand continues to rise. Smart energy technology is the soul of almost all smart city components such as vehicles, sensors, IoT devices, digital road signs, vehicles, parking meters, servers, stations, kiosk, etc. These components require power for actuation and operation accordingly to fulfill their potential, and as such there are no smart cities without smart energy technology to meet the needs of the city and its citizens.

To achieve smart energy technology in smart cities, the IoT, Big Data and ICT frameworks play a crucial role in acquiring the energy generation, storage, distribution, consumption, and waste data, analyzing the appropriate collected data, and making informed decisions that efficiently drive the overall smart energy technology infrastructure in meeting the energy needs of cities and citizens. The main components in smart energy involve having an advanced usage of technology, efficient energy transference, maintainable energy consumption, and an overall clean environment to ensure improvement in citizens' quality of life and a non-negative impact on the environment [2, 27, 28]. As smart cities invest in systems that enable a more robust and resilient smart energy infrastructure that addresses and provides optimal solutions to the energy needs of cities in both the short and long term, there are solutions that help address the energy needs in smart cities such as the development of renewable energy sources (RES) for the generation of power for many other systems, technologies, and applications in smart cities.

Many investments are made in RES that enable the potential to provide enough power to meet the power demands of cities, but these do not address the security issues and privacy implications that the overall smart energy technology generates. There are several inherited security related issues from IoT, Big Data and ICT frameworks involved, and as such the 3D privacy framework recommends addressing the privacy issues of the technologies and their components before deploying them in smart cities by putting in place needed regulations that would lessen privacy concerns of citizens to facilitate creating privacy aware smart cities. There are security concerns in the tracking and distribution of energy within smart cities to households and individuals, where the energy consumed is constantly tracked and measured.

Many privacy issues of control arise in associating the consumption of energy to individuals' presence in the homes with the notion of more individuals in the household would normally consume more energy. Thus, associating energy consumption times of the day and correlating it with the presence of individuals in the household becomes very worrying especially when the information pertaining to the energy consumption is directly associated with households' owner or renter in the smart energy ecosystem. The preference of the individual to choose a specific type of energy source from RES, and the need for individuals to generate off-grid power and contribute some of the energy to the grid with information pertaining to which households contribute what amount of power, create room for privacy concerns especially when IoT, Big Data and ICT frameworks are involved in facilitating the services [29, 30].

Due to the deployment and development of many off-grid contributions to the grid in terms of power, the sensitive information that is shared between the different energy providers in the network likewise produces perilous privacy concerns which can be intensified in an event of a breach or cyber threat attacks [29]. The 3D Privacy Framework recommends to appropriately evaluate smart energy applications, systems, and processes in smart cities to control and address privacy susceptibilities that pertain to what data is being collected from households and how to handle the collected data to ensure that privacy concerns are lessen through regulations wherever possible.

The 3D framework categorizes smart energy technology as it pertains to the city in quadrant 7, and as such it is associated with a medium privacy risk that imposes the adoption of guidelines in data collection and protocol to address security issues before more sensitive data are collected [1]. Table 2 reveals a summary of some rampant security induced privacy issues together with examples in literature pertaining to smart energy technology in smart cities that would have been prevented if IoT, Big Data and ICT frameworks enabling them were meticulously assessed and regulated to prevent excessive collection of personal identifiable data prior to being deployed in smart cities.

3.3 Smart Health Technology

Smart health technology in smart cities is a big contributor that enables many other smart technologies and applications to ensure citizens' quality of life and wellbeing continually improve. The growing number of citizens in smart cities demands a cautious consideration of the health state of citizens both in the short and long term as they undergo and adopt the different changes that take place in smart cities due new technologies, applications, and systems that they interact with regularly to fully benefit by living in smart cities. Smart cities today are filled with a myriad of health related IoT devices, applications and systems that leverage the use of Big Data and ICT infrastructures to enable different services within the smart city.

Smart health technology is at the center of the future of all smart cities because the health and the wellbeing of citizens are paramount if smart cities are to stand the test of time and meet the growing concerns of people migration toward them. There are no opportunities too appealing that would justify citizens to overlook their health and care as the Covid-19 pandemic has shown. It is evident that there are many smart health IoT components deployed in smart cities such as medical wearable devices, medical sensors, phone apps, smart watches, health trackers and medical bands, ECG, blood glucose, EMG, blood pressure, heart rate, gyroscope, accelerometers, and other motion sensors devices, etc. that present several inherited security concerns

Table 2 Security induced privacy concerns pertaining to smart energy technology in smart cities

Security issues	Induced privacy concerns	Threat examples	Possible resolution	References
Smart meters and power theft attacks	Authentication, authorization	Power theft, data manipulation, meter controlled by attackers	Anonymization, pseudonym, attribute-based credentials, message authentication codes, stealth operation	[31, 8, 9, 29–32]
Active cyber-attacks and terrorism	Authentication, authorization, confidentiality	US power system attack, spy grid penetration, attacker device control	Public-key infrastructure, dynamic encryption, code obfuscation, homomorphic encryption, Meter-based anti-viruses, usage loggers and SM rootkits	[33, 34, 32–36]
Active operational issues	Authentication, confidentiality	Systems data exposure issues, lack of self-regulation and reconfiguration, man in the middle attacks	Quality of Service (QoS) routing, Pre-programed self-healing action, smart grid communication program	[37, 30–32, 14]
Active secure Big Data, machine learning and artificial intelligence issues	Confidentiality, authentication	Model identification attacks, data intoxication attacks	Logical interface analysis, real-time system monitoring, power system model analysis, computer memory tagging	[37, 38, 13, 14, 30]
Big Data analytics attacks	Authentication, authorization	Data intoxication attacks	Incremental Harsh function	[39, 8, 12, 30]
Active digital fraud attacks	Confidentiality	Advanced metering infrastructure attacks, pricing attacks, data penetration and record attacks	Signature-based authentication, code obfuscation	[8, 9, 36, 14, 30]

for citizens, and generate some security induced privacy concerns when considered in the lens of enabling privacy aware smart cities.

All these components generate data that pertains to individuals, and as the data is shared through several avenues from system to system or device with different security levels, it provides an avenue for the data to be compromised especially when hacked or breached. The accomplishment of the state of art in smart health technology in smart cities relies on IoT, Big Data and ICT frameworks that play a major role in acquiring relevant health data and facilitating the development of health systems and avenues that respond to the urgent needs of citizens' healthcare ranging from temperature monitoring in babies to vital signs tracking in elderly [2, 40]. There are many smart health opportunities in almost all the sectors of human development that data acquisition could benefit with proper analysis and decision making.

The privacy aware smart city needs to open and develop more smart health solutions that better citizens' quality of life and give citizens avenues to connect to their health status anytime with real time applications without generating privacy concerns. As smart cities continue to invest and encourage initiatives in reliable smart health systems that enable a robust and resilient smart health infrastructure to provide optimal solutions to both the long and short term health needs of citizens, there are developments and investments in solution initiatives such as health embedded systems, machine learning, artificial intelligence and cloud computing that help address some individual health concerns among citizens as well as creating and deriving patterns from data to reflect the overall health trends in smart cities [40, 41].

As more health sensors are deployed to collect individuals' health data and information, the more security risks pertaining to IoT devices are experienced and can provide a gateway for malicious intents to compromise other sensitive data in the system. For accurate tracking and assessment of an individual's health, there is a need to identify, authenticate and associate a device with an individual's personal information to provide a better service, but if there are security incidents then privacy concerns may quickly arise. There are still several inherited security related issues pertaining to IoT, Big Data and ICT frameworks and as such the 3D privacy framework recommends in the case of health systems to address the privacy issues of the technologies and their components before deploying them in smart cities by putting in place proper regulations that lessen the privacy concerns of citizens to facilitate enabling privacy aware smart cities.

There are several security concerns in the collection of personal data to be linked with a smart health device, the connectivity technologies used, the health management systems accessed and the end-user interactions queries, such as adding and updating information [2]. The security issues in health systems give rise to privacy concerns in smart cities as most health systems pertaining to an individual user require some sort of identification and verification for data collection or retrieval in the application, and as such the individual's health information are communicated using available connectivity protocols such as Wi-Fi, NFC, BLE, ZigBee and Satellites using a mobile device where pseudonyms [39] and anonymization [9] combined with blockchain and encryption [32, 35] can be used to preserve the identity of the user, yet it may be harder to correctly associate and link data to the correct individual.

There is a big risk as it pertains to the health of citizens since it may be dangerous to, e.g., prescribe the wrong medicine to a patient as one example. Other privacy issues arise in considering on-body sensors that collect and send health information automatically using connectivity protocols that might be breached into and give access to how the health information is linked to personal information attached to on-body sensors [2]. There are many personalized health monitoring IoT devices that are used and linked to other IoT devices which share the same communication media as Bluetooth, SMS (short-messaging services), Wi-Fi, NFC (near field communication), and Social media where data can be breached and compromised due to insecure internet or connectivity protocols regarding data traffic loads [2].

The 3D Privacy Framework recommends to properly assess smart health technology systems in smart cities to better to regulate privacy vulnerabilities as security challenges involved in securing a smart healthcare system are dynamic and contain critical personal information that can be very damaging if tampered with for malicious intent [1]. The 3D Privacy Framework classifies smart energy technology in quadrant 2 and associates it to a high privacy risk space because there is potential of more personal information involved in providing health services and enabled quick and efficient solutions [1]. Table 3 displays a summary of some ubiquitous security induced privacy issues together with examples in the literature pertaining to smart health technology in smart cities that would have been prevented should IoT, Big Data and ICT frameworks enabling them were meticulously assessed and regulated to prevent excessive collection of personal identifiable data prior to their deployment in smart cities.

3.4   Smart Governance Technology

Smart governance technology in smart cities is the umbrella that covers the other smart technologies in terms of ensuring smart cities meet the goals they are created for toward the satisfaction of its citizens. It seems to be the most essential of the smart technologies because it is what makes the leaders of the cities justify to citizens why the investment in the deployed systems is essential to solve the problems of citizens and enable efficient and optimal use of resources. The growing number of citizens in smart cities imposes a wary reflection of the interaction between the government, citizens, and stakeholders coexisting together in the process of enabling smart cities to operate to their full potential [27]. Smart governance encompasses the use of IoT, Big Data and ICT in improving legal government decision making to facilitate the collaboration and adaptation of both smart cities' government and citizens to smart cities systems and technologies.

Smart governance in the context of smart cities can be demarcated as the potential of smart city government to utilize intelligent and adaptive resources at its disposal to enable better decision making about things and processes for the benefit of citizens

Table 3 Security induced privacy concerns pertaining to smart health technologies in smart cities

Security issues	Induced privacy concerns	Threat examples	Possible resolution	References
Active location based issues	Authentication, authorization	Denial of service with right credentials attack, data intoxication, FALSE diagnostics, and treatment attack	Pseudonym, attribute-based credentials, message authentication codes	[42–44]
Secure ICT Fog and cloud computing attack	Authentication, authorization	Fabricated node identity attacks	Cryptography, tamper-proofing via blockchain, data dissemination via FoG, M2M messaging with rule-based beacon, attribute based access control	[43–46]
Secure communication network issues	Authorization, authentication, confidentiality	Jamming communication channels, Sybil attacks in communication Man in the middle attack, replay attack, eavesdropping, blind letter attack, routing attack	Attribute-based credentials, location cloaking, homomorphic encryption, challenge-response protocol, steganography	[40, 42, 47, 63]
Big Data machine learning and artificial intelligence issues	Confidentiality, authentication	Fake emergency call attacks, data intoxication attacks	Signature-based authentication	[48, 5, 49]
Big Data analytics issues	Authentication	Insider and outsider attack, data intoxication attacks	Digital signatures, challenge-response protocol	[42, 48, 5, 49]
Passive secure controllers issues	Confidentiality	Dynamic inference attacks	Signature-based Authentication Access-based authentication, message authentication codes	[41, 47, 50]

to ensure efficiency and better quality of life in both the short and long term [28]. It is imperative for smart governance in smart cities to be successful because the success of smart cities lies in the way that citizens' quality of life and wellbeing are experienced. Smart governance technology possesses several components such as data, government, citizen, environment, and technology that enable the smart governance's ecosystem, which includes ICT-based tools such as social media, to help provide solutions to the numerous challenges requiring better governance like transparency, openness, democratization, citizen engagement, and information sharing.

The presence of IoT, Big Data and ICT frameworks is apparent in smart governance technology, and as such there are opportunities of security glitches that may lead to privacy concerns that should be addressed in smart cities to ensure they become more privacy aware smart cities. As data and information flow in the smart governance ecosystem, there is a great need to fully understand and distinguished the flow of personal and impersonal data as there is a greater chance of overlap in the services provided where it might be hard for example to separate the opinion of an individual with what they stand for and who they identify with [2, 28]. The role of technology in this space is crucial in enabling the optimization of urban services by providing extra avenues for feedback and interaction with citizens.

As smart cities invest in systems that enable more transparent and resilient smart governance technology that addresses and provides optimal governing solution to the citizens' needs in both the short and long term, there are solutions today that help address the governance's needs in smart cities in different ways such the electronic government (e-government), and the electronic governance (e-governance) that are classified in the highest privacy quadrant of the 3D privacy framework [1], namely in quadrant 3 and 4 of the framework respectively. There are countless inherited security related issues with both e-government and e-governance smart approaches as IoT, Big Data and ICT frameworks are involved in enabling the realization of connectivity and the interaction between the government and its citizens as citizen engagement voices citizens' views in smart cities [2, 4, 51].

The 3D privacy framework endorses assessing privacy issues of the used governance technologies, applications, and their associated components before deploying them in smart cities with proper regulations that lessen the privacy concerns of citizens to ease having privacy aware smart cities. There are privacy concerns in addressing anonymously the need for citizens participation without proper proof that they are indeed citizens of a particular smart city and not visitors [1]. The data collection process to verify that a citizen belongs to a smart city and provide feedback to the municipality through created e-government applications poses security issues as sensitive personal information are exchanged and can be breached. Sentiment monitoring likewise in smart cities through social media also poses similar privacy concerns [1, 52]. Avenues for democracy, social inequality and justice solutions also require some form of identification and authentication which can be jeopardized and affect the privacy of citizens [1].

The 3D Privacy Framework endorses the assessment of technologies in smart cities as they pertain to smart governance technology's regulation of privacy vulnerabilities regarding what data is being collected from citizens and how to manage the gathered data to ensure privacy concerns are lessened through regulations among citizens wherever possible [1]. The 3D framework classifies smart government and governance as they pertain to the city and citizen in quadrant 3 and 4 respectively [1], and as such they are associated with the highest privacy risks that demand the implementation of standards in data collection and provide an opportunity to deal with security issues before more sensitive data is compromised. Table 4 below highlights a summary of some extensive security induced privacy issues together with examples in literature pertaining to smart governance technology in smart cities that would have been prevented should IoT, Big Data and ICT frameworks enabling them to be meticulously assessed and regulated to prevent excessive collection of personal identifiable data prior to their deployment in smart cities.

It is vital to identify how citizens' privacy in smart cities can be jeopardized by smart applications, systems and processes deployed to enable smart technologies that benefit citizens yet tamper with their privacy. Regulations can be implemented to ensure that current and future innovations deployed do not continue to expose and criminalize citizens' privacy.

4 Conclusion

It is important to recognize how the same IoT, Big Data and ICT frameworks related security risks transcend between different technologies and leave loopholes that must be addressed to avoid creating more privacy concerns for citizens in smart cities. It is evident that privacy concerns of authentication, identification, authorization, and confidentiality are prevalent in almost all the presented technologies along with similar solutions that are identified in the various tables. Therefore, it is necessary to adopt the 3D privacy framework recommendations to strive for the creation of regulations and policies pertaining to each individual technology, application, system, and process in smart cities that define and determine how the privacy of citizens benefiting from the technologies is preserved throughout the lifespan of the service and beyond. This practice will proactively help protect citizens' privacy long before they reap the benefit of the enabled services, reducing the chances of attacks toward their privacy. Thus, by thoroughly addressing security issues pertaining to IoT, Big Data and ICT frameworks in smart cities, and providing regulations that encourage the preservation of the privacy of citizens, more privacy aware smart cities can come to fruition.

Table 4 Security induced privacy concerns pertaining to smart governance technologies in smart cities

Security issues	Induced privacy concerns	Threat examples	Possible resolution	References
IoT security issues	Authentication, authorization	Denial of service attack, remote exploitation, sensor-controlled failure, data intoxication, anonymity attacks	Pseudonym, Anonymization, attribute-based Credentials, message authentication approach	[53–57]
Secure ICT Fog and cloud computing attack	Authentication, authorization	Data leakage, Malicious insider attack, insecure API, malware injection attacks, digital disenfranchisement	Cryptography, blockchain, encryption	[53, 54, 56, 58]
Secure communication and social media network issues	Authorization, authentication, confidentiality	Impersonation attack, insecure protocol, third parties' attacks, lasting trust issues	Attribute-based credentials, location cloaking, homomorphic encryption, challenge-response protocol, Steganography	[55, 59–61]
Big Data machine learning and artificial intelligence algorithms issues	Confidentiality, authentication	Data intoxication attacks, bias algorithms, profiling location-based issues	Signature-based authentication	[53, 60–62]
Big Data analytics biasing issues	Authentication	Insider and outsider attack, sentiment misinterpretation, opinions versus facts dilemma	Digital signatures	[53, 59, 62, 61]

Acknowledgements The authors thank Arizona State University and the National Science Foundation for their funding support under Grant No. 1828010.

References

1. Mimo EM, Mcdaniel T (2021). 3D Privacy Framework: The Citizen Value Driven Privacy Framework. In 2021 IEEE International Smart Cities Conference, ISC2 2021 (2021 IEEE International Smart Cities Conference, ISC2 2021). Inst of Electr Electron Eng Inc. https://doi.org/10.1109/ISC253183.2021.9562841
2. Mohanty S (2016) Everything you wanted to know about smart cities. IEEE Consum Electron Mag 5:60–70. https://doi.org/10.1109/MCE.2016.2556879
3. Musafiri Mimo E, McDaniel T (2021) Security concerns and citizens' privacy implications in smart multimedia applications. In: Smart multimedia 2021: not yet published
4. Hahn D, Munir A Behzadan V (2019) Security and privacy issues in intelligent transportation systems: classification and challenges. IEEE intelligent transportation systems magazine. pp 1–1. https://doi.org/10.1109/MITS.2019.2898973
5. Serrano W (2021) Big Data in smart infrastructure. https://doi.org/10.1007/978-3-030-55187-2_51
6. Sehgal A, Perelman V, Kuryla S, Schonwalder J (2012) Management of resource constrained devices in the Internet of Things. IEEE Commun Mag 50(12):144–149
7. Koscher K, Czeskis A, Roesner F, Patel S, Kohno T, Checkoway S, McCoy D, Kantor B, Anderson D, Shacham H, Savage S, Experimental Security analysis of a modern automobile. In: 2010 IEEE symposium on security and privacy. IEEE, Berkeley, CA, USA, May 2010, pp 447–462
8. Chen H, Yang Y (2018) A practical scheme of smart grid privacy protection. IOP Conf Ser: Mater Sci Eng 394:042058. https://doi.org/10.1088/1757-899X/394/4/042058
9. Hoque S, Rahim A, Cerbo F (2014) Smart grid data anonymization for smart grid privacy. 470. https://doi.org/10.1007/978-3-319-12574-9_8
10. Mejri MN, Ben-Othman J, Hamdi M (2014) Survey on VANET security challenges and possible cryptographic solutions. Veh Commun 1(2):53–66
11. Calandriello G, Papadimitratos P, Hubaux J-P, Lioy A, Efficient and robust pseudonymous authentication in VANET. In: Proceedings of the fourth ACM international workshop on vehicular ad hoc networks. ACM, Montreal, Quebec, Canada, Sept 2007, pp 19–28
12. Yip S, Wong K, Phan RC, Tan S, Ku I, Hew W (2014) A privacy-preserving and cheat-resilient electricity consumption reporting scheme for smart grids. In: 2014 International conference on computer, information and telecommunication systems (CITS), 2014, pp 1–5. https://doi.org/10.1109/CITS.2014.6878971
13. Lee A, Brewer T (2009) Smart grid cyber security strategy and requirements. US Department of Commerce, Draft Interagency Technical Report NISTIR 7628. National Institute of Standards and Technology (NIST): Gaithersburg, MD
14. Otuoze A, Mustafa M, Larik RM (2018) Smart grids security challenges: classification by sources of threats 5:468–483. https://doi.org/10.1016/j.jesit.2018.01.001
15. Fuentes JMD, Gonz'alez-Tablas AI, Ribagorda A (2010) Overview of security issues in vehicular ad-hoc networks
16. Engoulou RG, Bella¨ıche M, Pierre S, Quintero A (2014) VANET security surveys. Comput Commun 44:1–13
17. Poudel B, Munir A (2017) Design and evaluation of a novel ECU architecture for secure and dependable automotive CPS. In: 2017 14th IEEE Annual consumer communications networking conference (CCNC). IEEE, Las Vegas, NV, USA, January 2017, pp 841–847

18. Behzadan V, Munir A (2017) Models and framework for adversarial attacks on complex adaptive systems. arXiv preprint arXiv:1709.04137
19. Kenney JB (2011) Dedicated short-range communications (DSRC) standards in the United States. Proc IEEE 99(7):1162–1182
20. He L, Zhu WT, Mitigating DoS attacks against signature-based authentication in VANETs. In: 2012 IEEE international conference on computer science and automation engineering (CSAE). IEEE, Zhangjiajie, China, May 2012, pp 261–265
21. Zhang Y, Pei Q, Dai F, Zhang L (2017) Efficient secure and privacy-preserving route reporting scheme for VANETs. J Phys: Conf Ser 910(1):012070
22. Christin D (2016) Privacy in mobile participatory sensing: current trends and future challenges. J Syst Softw 116:57–68
23. Yigitoglu E, Damiani ML, Abul O, Silvestri C, Privacy preserving sharing of sensitive semantic locations under road network constraints. In: 2012 IEEE 13th international conference on mobile data management. IEEE, July 2012, pp 186–195
24. Sakiz F, Sen S (2017) A survey of attacks and detection mechanisms on intelligent transportation systems: VANETs and IoV. Ad Hoc Netw 61:33–50
25. Poudel B, Giri NK, Munir A (2017) Design and comparative evaluation of GPGPU- and FPGA-based MPSoC ECU architectures for secure, dependable, and real-time automotive CPS. In: IEEE 28th International conference on application-specific systems, architectures and processors (ASAP). Seattle, WA, USA: IEEE, July 2017, pp 29–36
26. Oracevic A, Dilek S, Ozdemir S, Security in Internet of Things: A survey. In: 2017 International symposium on networks, computers and communications (ISNCC). IEEE, Marrakech, Morocco, May 2017, pp 1–6
27. Pereira G, Parycek P, Falco E, Kleinhans R (2018) Smart governance in the context of smart cities: a literature review. Inf Polity 23:1–20. https://doi.org/10.3233/IP-170067
28. Scholl HJ, AlAwadhi S (2016) Smart governance as key to multi-jurisdictional smart city initiatives: the case of the eCityGov alliance. Soc Sci Inf 55(2):255–277. https://doi.org/10.1177/0539018416629230
29. Goel S, Hong Y (2015) Security challenges in smart grid implementation. In: Smart grid security. Springer Briefs in Cybersecurity. Springer, London. https://doi.org/10.1007/978-1-4471-6663-4_1
30. Mendel J (2017). Smart grid cyber security challenges: overview and classification. e-mentor. 2017:55–66. https://doi.org/10.15219/em68.1282
31. Amin M (2008). Challenges in reliability, security, efficiency, and resilience of energy infrastructure: toward smart self-healing electric power grid, pp 1–5. https://doi.org/10.1109/PES.2008.4596791
32. Dias L, Rizzetti TA (2021) A review of privacy-preserving aggregation schemes for smart grid. IEEE Lat Am Trans 19(7):1109–1120. https://doi.org/10.1109/TLA.2021.9461839
33. Hasan MD, Mouftah HT (2015) Encryption as a service for smart grid advanced metering infrastructure, pp 216–221. https://doi.org/10.1109/ISCC.2015.7405519
34. Klimburg A (2011) Mobilising cyber power. Survival. 53. https://doi.org/10.1080/00396338.2011.555595
35. Singh P, Masud M, Hossain MS, Kaur A (2021) Blockchain and homomorphic encryption-based privacy-preserving data aggregation model in smart grid. Comput Electr Eng 93. https://doi.org/10.1016/j.compeleceng.2021.107209
36. Syed D, Refaat SS, Bouhali O (2020) Privacy preservation of data-driven models in smart grids using homomorphic encryption. Information. 11:357. https://doi.org/10.3390/info11070357
37. Staff CACM (2016) Future cyberdefenses will defeat cyberattacks on PCs. Commun ACM 59:8–9. https://doi.org/10.1145/2963167
38. Yu X, Xue Y (2016) Smart grids: a cyber-physical systems perspective. Proc IEEE 104:1–13. https://doi.org/10.1109/JPROC.2015.2503119
39. Murphy MH (2018) Pseudonymisation and the smart city: considering the general data protection regulation. https://doi.org/10.4324/9781351182409-14

40. Arca S, Hewett R (2020) Privacy protection in smart health, pp 1–8. https://doi.org/10.1145/3406601.3406620
41. Jagadeesh R, Mahantesh (2019) Privacy and security issues in smart health care, pp 378–383. https://doi.org/10.1109/ICEECCOT46775.2019.9114681
42. El-Bakkouri N, Mazri T (2020) Security threats in smart healthcare. ISPRS—Int Arch Photogrammetry, Remote Sens Spatial Inf Sci. XLIV-4/W3-2020:209–214. https://doi.org/10.5194/isprs-archives-XLIV-4-W3-2020-209-2020
43. Ullah A, Sehr I, Akbar M, Ning H (2018) FoG assisted secure de-duplicated data dissemination in smart healthcare IoT, pp 166–171. https://doi.org/10.1109/SmartIoT.2018.00038
44. Zhang Y, Zheng D, Deng R (2018) Security and privacy in smart health: efficient policy-hiding attribute-based access control. IEEE Internet Things J 5:2130–2145. https://doi.org/10.1109/JIOT.2018.2825289
45. Cao S, Zhang G, Liu P, Zhang X, Neri F (2019) Cloud-assisted secure eHealth systems for tamper-proofing EHR via blockchain. Inf Sci. 485. https://doi.org/10.1016/j.ins.2019.02.038
46. Saguil D, Xue Q, Mahmoud Q (2019) M2MHub: a blockchain-based approach for tracking M2M message provenance, pp 1–8. https://doi.org/10.1109/AICCSA47632.2019.9035210
47. Reddy B, Adilakshmi T (2021) Distributed and decentralized attribute based access control for smart health care data. https://doi.org/10.1007/978-981-16-1502-3_8
48. Rabhi L, Noureddine F, Afraites L, Belaid B (2018) Big Data analytics for smart health state of the art and issues
49. Voros S, Moreau-gaudry A (2014) Sensor, signal, and imaging informatics: Big Data and smart health technologies. Yearb Med Inf 9:150–153. https://doi.org/10.15265/IY-2014-0035
50. Ding Y, Sato H (2020) Derepo: a distributed privacy-preserving data repository with decentralized access control for smart health. https://doi.org/10.1109/CSCloud-EdgeCom49738.2020.00015
51. Sookhak M, Tang H, He Y, Yu FR (2019) Security and privacy of smart cities: a survey, research issues and challenges. In: IEEE Commun Surv Tutorials 21(2):1718–1743. Secondquarter 2019. https://doi.org/10.1109/COMST.2018.2867288
52. Zoonen L (2016) Privacy concerns in smart cities. Gov Inf Quart 33. https://doi.org/10.1016/j.giq.2016.06.004
53. Mora H, Pujol F, Morales M, Mollá Sirvent R (2021) Disruptive technologies for enabling smart government. https://doi.org/10.1007/978-3-030-62066-0_6
54. Ahmed E (2020) The impact of blockchain in smart government services. https://doi.org/10.31220/osf.io/hazmb
55. Guida P (2021) PNRR e smart governance. Project Manager (IL). 4–5. https://doi.org/10.3280/PM2021-046001
56. Rahmah M, Ameen A, Isaac O, Al-Shibami A, Bhumik A (2020) Impact of smart government usage and smart government effectiveness on employee happiness. 82:12086–12100
57. Abosaq NH (2019) Impact of privacy issues on smart city services in a model smart city. Int J Adv Comput Sci Appl 10(2):177–185
58. Abi Sen AA, Eassa FA, Jambi K (2018) Preserving privacy of smart cities based on the fog computing. https://doi.org/10.1007/978-3-319-94180-6_18
59. Ismagilova E, Hughes L, Rana NP et al (2020) Security, privacy and risks within smart cities: literature review and development of a smart city interaction framework. Inf Syst Front. https://doi.org/10.1007/s10796-020-10044-1
60. Alryalat M, Rana NP, Dwivedi YK (2015) Citizen's adoption of an e-government system: validating the extended theory of reasoned action (TRA). Int J Electron Gov Res 11(4):1–23
61. Alter S (2019) Making sense of smartness in the context of smart devices and smart systems. Inf Syst Front, pp 1–13. https://doi.org/10.1007/s10796-019-09919-9
62. Yahia NB, Eljaoued W, Saoud NBB, Colomo-Palacios R (2019) Towards sustainable collaborative networks for smart cities co-governance. Int J Inf Manage. https://doi.org/10.1016/j.ijinfomgt.2019.11.005
63. Ranjith J, Mahantesh K (2019) Privacy and security issues in smart health care. In: 2019 4th International conference on electrical, electronics, communication, computer technologies and

optimization techniques (ICEECCOT), 2019, pp 378–383. https://doi.org/10.1109/ICEECC OT46775.2019.9114681

64. Paiva S, Ahad M, Zafar S, Tripathi G, Khalique A, Hussain I (2020) Privacy and security challenges in smart and sustainable mobility. SN Appl Sc. 2. https://doi.org/10.1007/s42452-020-2984-9

65. Samih H (2019) Smart cities and internet of things. J Inf Technol Case Appl Res 21:1–10. https://doi.org/10.1080/15228053.2019.1587572

Ethics of Face Recognition in Smart Cities Toward Trustworthy AI

Mengjun Tao, Richard Jiang, and Carolyn Downs

Abstract In the past few decades, thanks to the continuous development of machine learning and deep learning algorithms, as well as the continuous improvement of computing power and databases, facial recognition technology (FRT) has developed rapidly. Widespread use of this technology can be seen in all fields of life, such as facepay, individual identification, smart-city surveillance, e-passport or even face to DNA identification. However, some experts believe that certain errors that commonly creep into its functionality and a few ethical considerations need to be addressed before its most elaborate applications can be realized. This article explores the ethical issues of FRT used in different scenarios, tries to examine some legal and regulatory issues that may be encountered during the use of FRT, and technically analyze how to build a trustworthy intelligent system.

Keywords Face Recognition · AI ethics · Legislation

1 Introduction

Facial recognition technology (FRT) has now been widely used in many applications. Some unnoticeable places in city streets may have many cameras with FRT; facial recognition payment tools; facial recognition equipment encountered in various identification places; police Facial recognition aids used when arresting suspects; face identification or verification used with DNA information when verifying biological materials, etc. In these cases, a lot of user privacy is used. However, whenever it comes to the sensitive topic of user privacy, individuals are always concerned about whether the government, enterprise, or scientific research institution's use of user privacy is standardized and reasonable. And in many countries, researchers, as well as civil-liberties advocates and legal scholars, are among those disturbed by facial recognition's rise, people even resist the rise of facial recognition. Opponents are also concerned that police and law-enforcement agencies are using FRT

M. Tao (✉) · R. Jiang · C. Downs
LIRA Center, Lancaster University, Lancaster LA1 4YW, UK
e-mail: m.tao2@lancaster.ac.uk

© Springer Nature Switzerland AG 2022
R. Jiang et al. (eds.), *Big Data Privacy and Security in Smart Cities*,
Advanced Sciences and Technologies for Security Applications,
https://doi.org/10.1007/978-3-031-04424-3_2

in discriminatory ways, and that governments could employ it to repress opposition, target protesters or otherwise limit freedoms [1]. Meanwhile, some ethicists state that Scientists should acknowledge the morally dubious foundations of much of the academic work in the field—including studies that have collected enormous data sets of images of people's faces without consent, many of which helped hone commercial or military surveillance algorithms [2]. Like other AI-powered technologies, facial recognition systems need to follow a few ethical principles while being used for various purposes. In addition to the protection of user privacy, there are other ethical issues that require our attention. In 2018, a seminal paper by computer scientists Timnit Gebru, then at Microsoft Research in New York City and now at Google in Mountain View, California, and Joy Buolamwini at the Massachusetts Institute of Technology in Cambridge found that leading facial-recognition software packages performed much worse at identifying the gender of women and people of color than at classifying male, white faces [3]. This case also raised people's concerns about the difference in the accuracy of face recognition caused by demographic differences.

The remaining articles will be discussed in the following format.

In Sect. 2, we started with the four representative aspects of Demographic aspect, Smart-city surveillance, Facepay and Face to DNA, and analyzed in detail a series of ethical issues that may arise when FRT is applied in these scenarios.

Although both have their own privacy protection regulations, there are still certain differences between the United States and Europe in terms of specific laws and regulations. In Sect. 3, we detailed some regulations that may be violated during the application of FRT.

In Sect. 4, Considering the ethical issues that may exist in the use of FRT mentioned earlier, we cut in from the five directions of Explainability, privacy, Nondiscrimination & Fairness, Integrity and environmental friendliness, and discussed some representative technical methods to realize trustworthy AI.

In the process of human development, there are endless problems encountered, and human values and social moral standards will also be constantly updated. Therefore, as an intelligent machine, it should also keep up with the pace of the times and continue to improve. This is what we are discussing in Sect. 5.

2 Representative Scene

2.1 Demographic Aspect

FRT applications are already well-known in police investigations. At least 14 UK police forces have made use of crime-prediction software or plan to do so, according to Liberty. It believes the programs involved can lead to biased policing strategies that unfairly focus on ethnic minorities and lower-income communities [4]. In fact, such prejudiced police cases did happen, and not just in the UK. According to a publication of the Penn Program on Regulation, After Detroit police arrested Robert

Williams for another person's crime, officers reportedly showed him the surveillance video image of another Black man that they had used to identify Williams. The image prompted him to ask the officers if they thought "all Black men look alike." Police falsely arrested Williams after facial recognition technology matched him to the image of a suspect—an image that Williams maintains did not look like him [5]. The police officers involved had trusted facial recognition AI to catch their man, but the tool hadn't learned how to recognize the differences between black faces because the images used to train it had mostly been of white faces [6]. Some experts realize that even artificial intelligence may have human bias, because even if the algorithms designed by experienced experts are superior, the data used to train these algorithms or models may be lopsided, which means data could reflect racial, gender, and other human biases. Luka Crnkovic-Friis, chief executive of AI start-up Peltarion and co-founder of the Swedish AI Council, told the BBC: "Because AI is trained by people, it's inevitable that bias will filter through." Automation tools are only ever as good as the data fed into them, so when using historical data where there is a strong human bias—such as race, re-offending rates and crime—there certainly is a risk that the results could produce bias and the government is right to take steps to account for this" [7]. The moment the artificial intelligence product was born, it was determined that it could not reflect the needs of everyone in the world, because the data that was trained at the time only represents the information of a part of people, and it cannot objectively reflect the situation of everyone, let alone said it applies to the world. In addition to the possible one-sidedness of data, some scientists may have some biases themselves, so it is not surprising that artificial intelligence products designed by them are biased. Like many disciplines, often those who perpetuate bias are doing it while attempting to come up with something better than before. However, the predominant thought that scientists are "objective" clouds them from being self-critical and analyzing what predominant discriminatory view of the day they could be encoding, or what goal they are helping advance [8] (Fig. 1).

The problem is not entirely new. Back in 1988, the UK Commission for Racial Equality found a British medical school guilty of discrimination. The computer program it was using to determine which applicants would be invited for interviews was determined to be biased against women and those with non-European names. However, the program had been developed to match human admissions decisions, doing so with 90–95% accuracy. What's more, the school had a higher proportion of non-European students admitted than most other London medical schools. Using an algorithm didn't cure biased human decision-making. But simply returning to human decision-makers would not solve the problem either [9]. How to solve the prejudice problem in artificial intelligence products has been plagued the government and scientists in recent years (Fig. 2).

In order to alleviate such problems, starting from the source, some experts have proposed that researchers should ensure that the data used to train the AI algorithm should reflects the various situations in the application scenario as much as possible. At the same time, researchers should also improve their professionalism, establish correct concepts, and avoid bias when coding. Andrew Ng, adjunct professor at Stanford University, an American computer scientist and technology

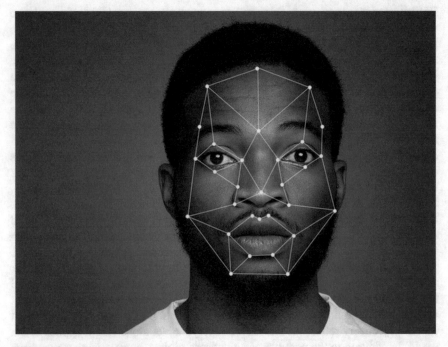

Fig. 1 FRT technology has repeatedly experienced serious lack of fairness in accuracy in places such as people of color and gender

entrepreneur focusing on machine learning and AI, gave some ways to reduce the bias/discrimination of AI systems: Expand the types of data; Use unbiased basic data; make the biased word "Return to 0"; make the audit more transparent. In addition to program developers, people also shoulder many important responsibilities from an enterprise perspective. According to an article in Harvard Business Review, the author proposes six essential steps for CEOs and their top management teams to do to lead the way of AI tech fairer [9]:

- business leaders will need to stay up to-date on this fast-moving field of research.
- when your business or organization is deploying AI, establish responsible processes that can mitigate bias.
- engage in fact-based conversations around potential human biases.
- consider how humans and machines can work together to mitigate bias.
- invest more, provide more data, and take a multi-disciplinary approach in bias research (while respecting privacy) to continue advancing this field.
- invest more in diversifying the AI field itself. A more diverse AI community would be better equipped to anticipate, review, and spot bias and engage communities affected [9]. The original article has a more detailed explanation, readers who are interested can refer to it.

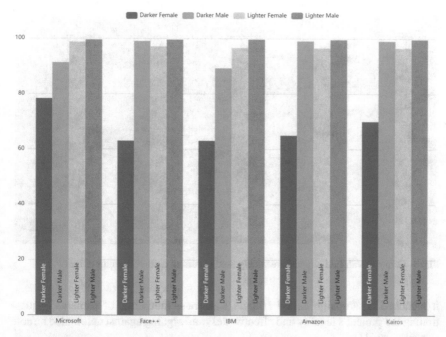

Fig. 2 The figure revealed discrepancies in the classification accuracy of face recognition technologies for different skin tones and sexes

2.2 Smart-City Surveillance

Compared with a few decades ago, the number of cameras on the streets is increasing rapidly. Maybe in some places you haven't noticed, your facial information has been scanned and analyzed by the overhead camera. Whether you like it or not, the fact is that cameras equipped with FRT are purchased by government agencies from third-party commercial companies and placed in the streets of citizens' daily lives (Fig. 3).

Nissenbaum's argument for privacy in public rests on two premises. First, citing empirical data, she claims that many people are dismayed when they learn that personal information is collected about them with- out their consent, even when they are in public places. This negative response shows that many people do indeed have some privacy expectations even when they are in public spaces. Second, she argues that these popular sentiments can be justified by analyzing characteristics of public data harvesting through electronic means that make it quite different from the everyday observation of people in public places [10]. In most cases, citizens have never participated in surveys about whether they agree to install cameras on the streets. For example, the government didn't ask Belgrade's residents whether they wanted the cameras, says Danilo Krivokapić, who directs a human-rights organization called the SHARE Foundation, based in the city's old town. This year,

Fig. 3 Cameras with FRT support are widely used in current cities to monitor citizens

it launched a campaign called Hiljade Kamera—'thousands of cameras'—questioning the project's legality and effectiveness and arguing against automated remote surveillance [1].

The digital revolution transforms people's view about values and priorities. Automated facial recognition (AFR) comes with many concerns as well as benefits. The technology raises significant legal and ethical challenges, which risk perpetuating systemic injustice unless countervailing measures are put in place. The way facial images are obtained and used, potentially without consent or opportunities to opt out, can have a negative impact on people's privacy [11]. In a smart city, there is tremendous data to deal with. Anyone who have the access to this data would have threat to citizens' privacy safe in theory. Despite the individuals, organizations or companies that have database management and use rights may also pose a certain threat to people's personal data privacy. In 2020, according to Reuters: Google was sued in a class action lawsuit on June 2 and accused of violating millions of users' privacy by using browsers set to incognito mode to track users' Internet usage extensively [12]. Even the Google with such a large brand influence on a global scale can make this kind of blunder, how can we trust other companies? Humans are not perfect. When decisions have to be made, a lot of factors usually come into the picture, some of which may lead to biased or discriminatory actions. The availability of this data increases the possibilities of misuse by those who have it. Observing ethical conduct becomes more necessary than ever. Rules, policies, regulations, and laws can limit misconduct, but there is always room for major problems given the vast amount of data available and the deep penetration of the data collection process in every aspect of residents' lives [13]. In addition to the threats that data managers may cause, whether data storage is secure enough to withstand technical intrusions by external hackers is also a question. If the database firewall itself is very fragile, it may often be exposed to external intrusions, and user data privacy will become very dangerous.

In fact, more and more people are aware of the potential harm that cameras bring to personal privacy, and residents in some areas have resolutely stood up to resist this type of urban surveillance behavior. On December 18, 2020, The New Orleans City Council passed a new law on Thursday that regulates certain parts of the city's surveillance system and places an outright ban on specific pieces of surveillance technology, including facial recognition software and predictive policing [14]. "Eye on Surveillance is proud of this important win for the people of New Orleans and looks forward to working with neighbors and residents to explore evidence-based options for public safety," a statement from the group said. "We will also continue to fight for further City action to address components of our original proposal that were not included in the final ordinance." Similar situation occurred in New York, the NYPD is facing multiple lawsuits over the use and disclosure of its digital surveillance practices. The New York City Police Department has eyes everywhere. Before the passage of the Public Oversight of Surveillance Technology Act (POST Act) by the New York City Council in June, the public had few ways to learn how the NYPD was using surveillance technologies like facial recognition, running checks against controversial databases or even tracking public opinion about the department itself. While those who pushed for the passage of the POST Act, including Gibson, Cahn and Díaz, said the legislation will create more transparency about the NYPD's use of systems like facial recognition as well as tools like drones and license plate readers—hopefully preventing the need for lawsuits seeking that information—they agreed that more work lies ahead in figuring out how the NYPD should or shouldn't be using these technologies [15]. Through the above information, it is not enough to rely on legislation to solve various problems caused by city surveillance. Fortunately, apart from the legislation, there are still some other solutions proposed by experts to alleviate the privacy threat due to mass surveillance. The general view is that it needs to involve ethical practices in some form, whether that be through market-based regulation, technological fixes, policy changes, governance mechanisms, legal interventions, or a more fundamental rethink of the logic and aims of smart cities [16]. These several aspects give people ideas for thinking about solutions. For instance, for market-based supervision, people can strengthen management and constraints on companies that provide FRT camera support. By signing a strict agreement, these companies ensure that the user's private data is reasonable and legal in the process of data storage, usage, transmission, and destruction. For technological fixes, equipment maintenance units should update and repair cameras and other equipment in a timely manner, so as not to give criminals a chance to take advantage of it, and to ensure that users' privacy will not be threatened by external forces. For governance mechanisms, the supervision of law enforcement agencies should be strengthened to avoid the abuse of power, which may lead to private data being called or tampered with. Citizens should also take the initiative to use the right to supervise, and for unjustifiable and illegal behaviors of enterprises or government departments, they should bravely stand up and point out the problem and refuse the recurrence of the problem.

2.3 Facepay

With the continuous development of the Internet industry, offline physical retail companies have undergone a huge technological revolution. And mobile payment plays a vital role in it. [17]. Facial biometrics or biometric facial recognition can determine a person's identity through her face. After this technology was applied to payment, a popular trend began to appear all over the world. The application is relatively wide, with a relatively large number of users, such as China's Alipay and WeChat Pay. When WeChat and Alipay both launched their own Facial Recognition Payment (FRP) systems "Frog" and "Dragonfly 2.0" in March and April 2019 respectively, China officially entered the era of facial payments [18] (Fig. 4 and Table 1).

Over the years, most of the consumers have also experienced the rapid development of face payment. In a country where mobile payments have spread by leaps and bounds in the past couple of years, some 1000 convenience stores have already installed a facial payment system, and more than 100 million Chinese have registered to use the technology [19]. Not only China, BUT the US has also developed the first facial payment method named PopID in California. However, in addition to China, facial payment has not been developed in other countries or regions as common as China. The US, for example, famously began piloting facial payments several years ago with the likes of Amazon Go stores. Yet, for all the hype surrounding

Fig. 4 Face payment is very common in the Chinese market. Customers only need to scan their face against a specific device to complete the payment when paying

Table 1 Laws, specifications and privacy policies related to FRP in China

Platforms' privacy policy	
Mobile payment platforms	Content related to FRP
Alipay	The Alipay Privacy Policy and General Rule for Biometric Recognitionh stipulate the details of facial payment services such as the storage and protection of facial information
Wechat pay	The TenPay Privacy Policy and the Service Agreement of FRP clarify service functions and privacy protection details
UnionPay app	The UnionPay Privacy Policy specifies the scope of personal information including facial images and the identity number

such projects, they still fall short of the capabilities and acceptance seen in China. To be fair though, the reason Amazon Go shoppers also need their mobile is likely less of a technical limitation than a compliance issue. In both the US and the EU there are exacting standards set for payments authentication. According to EU GDPR laws (which are generally like US laws) every payment must be authenticated by a combination of two of the three following touchpoints: (1) knowledge—something only the user knows; (2) possession—something only the user possesses; and (3) inherence—something the user is [20].

In fact, there is indeed a reason why the EU GDPR laws and the US CCPA laws impose such strict restrictions on payment authorization. Some experts proposed potential threat in facial payment. "At present, abuse of facial recognition may threaten an individual's payment security, but its potential risks may extend to threats against their personal safety, such as drone attacks using facial recognition," Liu Gang, director of the Nankai Institute of Economics and chief economist at the Chinese Institute of New Generation Artificial Intelligence Development Strategies, told the Global Times on Wednesday [21]. While consumers may enjoy the convenience that the new technology brings, they may also be confronted with serious threats to their portrait rights, personality rights and property rights [22]. Take PopID as an example, level of protection is baked into PopID's user agreement and basic structure. Customers choose to sign up for the system and must click a button or tell a cashier every time they use it, setting it apart from the kind of passive surveillance that most privacy advocates argue is ripe for abuse. PopID's software also runs on stand-alone devices, which means companies can't simply connect their own security cameras and start logging their employees' every move in a searchable database. Most important, the agreement signed by users when joining the service makes clear that PopID will share user data only when customers explicitly tell it to, whether that means pushing a button to pay or signing up for a loyalty points system with a given shop. Miller, the 42-year-old Pasadena entrepreneur who founded and runs PopID, said the company would treat law enforcement like any other third party. If the Los Angeles Police Department came to PopID and asked to run a photo against its database, "our answer to the LAPD would be that we are not allowed to share that information," Miller said. "We can't do it, sorry—this is a consumer opt-in service" [23]. As a result, third-party mobile companies such as Alipay and WeChat have issued a series of service agreements and general rules of service for biological recognition to further define the service functions, the scope of the company's responsibilities and the details of privacy protection [24].

In addition to strictly complying with privacy agreements, companies must also update and improve their own artificial intelligence algorithms and hardware to avoid loopholes that may be exploited by people with ulterior motives. As early as the early days of the face recognition algorithm, it appeared that using the user's photo in front of the camera could complete the face comparison and unlock it. In order to improve their own face recognition technology, different companies have adopted different advanced methods. Take Apple's Face ID as an example, Face ID provides intuitive and secure authentication enabled by the state-of-the-art TrueDepth camera system with advanced technologies to accurately map the geometry of individual's face. The TrueDepth camera captures accurate face data by projecting and analyzing thousands of invisible dots to create a depth map of individual's face and also captures an infrared image of face [25]. However, if attackers cannot spoof a system, they might try to force an authentication. For example, they could hold a person's phone to the owner's face when asleep or coerce them to unlock it [26]. For such a special duress situation, can algorithm developers add new recognition functions, such as automatically triggering the emergency call system by recognizing the pre-set facial expressions that indicate danger of the device owner. Potential customers expect to find what they need easily and immediately. This requires knowing what your customers want so you can incorporate interesting features and relevant content [27]. Therefore, face payment still has many potential security issues that need to be resolved.

In conclusion, if in the future, face-scanning payment is used on a large scale in the world, it will inevitably bring a large amount of user privacy data. Service providers have the responsibility and obligation to strictly abide by privacy agreements, protect customer data security, and improve algorithms and hardware support for users in time. In addition, they should also consider special circumstances, such as users receiving coercion to swipe their faces to open their devices for payment. Users should also pay attention to protecting their privacy and security when performing Facepay operations and prevent others from stealing their own information.

2.4   Face to DNA

In addition to the above-mentioned common face recognition technologies, there are still some more interesting and novel face recognition methods which compare the DNA information by inputting the face or searches the corresponding face image in the database through the DNA information. Our physical appearance, including our face, is hardwired into our genetic material. Scientists have already identified multiple genes that determine the shape of our face from the distance between our nostrils to the shape of our chin. Predicting what someone's face looks like based on a DNA sample remains a hard nut to crack for science. It is, however, getting easier

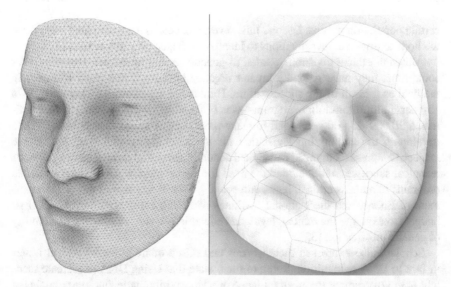

Fig. 5 Facial recognition from DNA refers to the identification or verification of unidentified biological material against facial images with known identity. One approach to establish the identity of unidentified biological material is to predict the face from DNA, and subsequently to match against facial images. However, DNA phenotyping of the human face remains challenging [48]

to use such a sample to filter the right face from a face database [28]. Furthermore, even if the corresponding face cannot be compared in the database, the expert can still build an artificial face through the characteristics of DNA (Fig. 5).

As we all know, specific fragments of DNA correspond to specific facial features, which is also the basic principles of face to DNA recognition. Applying GWAS to multivariate shape phenotypes, we identified 203 genomic regions associated with normal-range facial variation, 117 of which are novel [29]. They have identified more than 130 chromosomal regions associated with specific aspects of facial shape. Another group from departments of Oral Biology in University of Pittsburgh scanned the DNA of more than 8000 individuals to look for statistical relationships between about seven million genetic markers—known locations in the genetic code where humans vary—and dozens of shape measurements derived from 3D facial images [30]. DNA phenotyping has been an active area of research by academics for several years now. For example, Forensic biology researchers Manfred Kayser and Susan Walsh, among others, have pioneered several DNA phenotyping methods for forensics. In 2010, they developed the IrisPlex system, which uses six DNA markers to determine whether someone has blue or brown eyes. In 2012, additional markers were included to predict hair color. Last year the group added skin color [31]. Therefore, forensic DNA phenotyping provides more details about the subject to which a given biological sample belongs, without the need for a reference sample for comparative analysis. And also, some ethical and legal aspects should be taken into account so that this new technology does not promote segregation or ethnic persecution of certain population groups. Despite this, several real cases have benefited from these methods to orientate investigations to identify both suspects and victims [32].

Although ethical questions related to genetic testing have been recognized for some time, they have gained a greater urgency because of the rapid advances in the field as a result of the success of the Human Genome Project. That project—a 13-year multibillion-dollar program—was initiated in 1990 to identify all the estimated 20,000–25,000 genes and to make them accessible for further study. When developing the authorizing legislation for the federally funded Human Genome Project, Congress recognized that ethical conundrums would result from the project's technical successes and included the need for the development of federally funded programs to address ethical, legal, and social issues. Accordingly, the U.S. Department of Energy and the National Institutes of Health earmarked portions of their budgets to examine the ethical, legal, and social issues surrounding the availability of genetic information [33].

Scholars also expressed their worries and raised many potential ethical issues in face to DNA recognition. Some of them state that Using DNA to recreate faces will help government mass surveillance projects. While the technology remains at the primary stages of development, it can already generate rough facial reconstructions, good enough to be used to narrow a manhunt or even eliminate suspects from an investigation. This is what has raised red flags with experts who work in the ethical side of scientific study who are concerned about the downsides to allowing governments to have the power to interrogate people's DNA [34]. In addition to the government's possible greater monitoring of the people, companies that help the government to manage and analyze DNA data may also have potential problems. On 7th June 2018, according to media The Verge, DNA testing service company MyHeritage revealed this week that hackers have stolen the personal information of 92 million user accounts. The reason why hackers are interested in DNA information, some scholars have also given professional answers. Giovanni Vigna, a professor of computer science at the University of California, Santa Barbara and co-founder of the network security company Lastline, said that one simple reason is that hackers may want to get a ransom through DNA data. If the money is not paid, the hacker may threaten to revoke access or post sensitive information online; for this reason, an Indiana hospital paid the hacker $55,000. But genetic data can be profitable. "These data can be sold at low prices or sold to insurance companies through monetization," Vigna added. "You can imagine the consequences: one day, I might apply for a long-term loan and be rejected because there are data that show that before I repay the loan, I will probably get Alzheimer's disease and die." A selection-based model explains the observed difference between essential and disease genes and suggests that diseases caused by somatic mutations should not be peripheral, a prediction we confirm for cancer genes [34]. This shows that some diseases can be predicted through gene recognition. If citizens' genetic information is abused and insurance companies cooperate with companies in the field of genetic analysis, many people with underlying diseases will be refused insurance by insurance companies, and the insurance industry will be extremely unfair to ordinary people. This has aroused criticism, and for many people, the risk of discrimination is the most worrying issue. In the United States, the Genetic Information Non-Discrimination Act protects individuals from discrimination by employers and health insurance companies. However, it does not apply to life insurance and disability insurance, nor does it protect people from discrimination in other areas such as education and housing. In fact, there are already many people have been worried about the genetic information. Experts stated at the RSA Conference in San Francisco on 26th February 2020 that patients have given up genetic tests that may benefit their health due to concerns that genetic information may be used in unexpected ways. Patrick Courneya, chief medical officer of Kaiser Permanente, said that people have begun to abandon genetic testing because of privacy concerns. He did not specify what specific issues caused patients to avoid some tests, but some of them may stem from concerns about genetic genealogy. Genetic genealogy is a crime-solving technology that combines genetics and family history. Bioethicists and civil liberties watchdogs have raised concerns about the potential problems of this approach, because anyone who shares their DNA in the database will make decisions for those who share their genetic makeup.

In order to protect personal DNA and other private data, there is much that needs to be done. The Health Insurance Portability and Accountability Act (HIPAA) promulgated by the United States in 1996 is a data privacy and security clause for the protection of medical information. With the increasing number of cyber-attacks and health data breaches, this clause becomes more important. HIPPA has formulated a series of rules for personal medical information and protecting it from unauthorized use. With privacy and confidentiality as the two main points, it specifies the issues that medical institutions should pay attention to when handling patient information [35]. Experts have also put forward many practical methods:

- User anonymity: When purchasing genetic testing services, enabling pseudo-anonymous payments can eliminate several potential vulnerabilities faced by individuals. Most importantly, keeping individuals pseudo-anonymous will eliminate personal genome companies' reliance on data identification before data sharing. In addition, because there is no need to collect customer data, such as name and credit card information, the risk of individuals using this system being affected by security breaches is reduced.
- Data access control: Individuals should have complete control over their personal genome data. However, today DTC Genomics companies effectively own and control all the genomic data they produce. This brings several risks. First, the centralized genome database may be compromised by hackers, which has happened in the past. Second, government agencies can compulsorily access the genome database. Third, because there is no inspection in place, personal genome companies may intentionally or unintentionally violate data privacy. Delegated access control does not rely on one party, but multiple independent organizations, which may be a reasonable compromise between security and availability. To this end, the key used to encrypt genomic data can be divided into several parts and distributed to multiple independent parties.
- Record auditability: In order to build trust and encourage genomic data sharing, data access requests and user consent must be communicated transparently and

remain unchanged, which will ensure auditability and prevent abuse. This can be achieved using blockchain-an immutable public database maintained by a peer-to-peer network. Network participants can propose to add new entries to other participants in the blockchain network through broadcast transactions. Only after verification by most participants, the network will accept new transactions. Transactions are bundled into time-stamped blocks, and each block references its previous block. This creates an ordering and prevents deletion of data stored on the blockchain.

3   Relevant Laws or Regulations

Nowadays, face recognition technology has been ubiquitous in human society, and applications of this technology can be found in various scenarios in all walks of life. Personally, the technology can be used for facial recognition to open the phone and complete the payment. When you go home, you can swipe your face to open the door, etc. In terms of society, FRT is widely used in urban surveillance, public transportation and other fields. Some of the most worrying applications of the technology are in law enforcement, with police departments and other bodies in the United States, Europe, and elsewhere around the world using public and private databases of photos to identify criminal suspects and conduct real-time surveillance of public spaces [36]. In this regard, laws and regulations specific to this technology have also emerged in various places in attempt to alleviate citizens' worries and strengthen the supervision of this technology. Certain jurisdictions have imposed bans on its use while others have implemented more targeted interventions. In some cases, laws and regulations written to address other technologies may apply to facial recognition as well (Fig. 6).

Although the starting point is the same—in order to strengthen the supervision of FRT and protect user privacy data, there are still certain differences in laws and regulations of different countries and regions. In the case of the USA, it is one of the main global regions in which the technology is being rapidly evolved, and yet, it has a patchwork of legislation with less emphasis on data protection and privacy. Within the context of the EU and the UK, there has been a critical focus on the development of accountability requirements particularly when considered in the context of the EU's General Data Protection Regulation (GDPR) and the legal focus on Privacy by Design (PbD). However, globally, there is no standardised human rights framework and regulatory requirements that can be easily applied to FRT rollout [37]. Then this article will separately introduce the laws and regulations of the United States and European countries regarding FRT supervision and user privacy protection.

Fig. 6 The geographical scope of European privacy laws is being continuously expanded through General Data Protection Regulation (GDPR). And it is precisely because of this expansion of the scope of application that many organizations located outside the EU and not subject to the current European privacy laws will also have to apply the EU privacy laws with the implementation of the GDPR. The GDPR requires these organizations to respond to GDPR's "compliance" requirements in a timely manner

3.1 Related U.S. Laws and Regulations

Regarding face recognition and user privacy data, each state or city government and civil organizations in the United States have their own laws or regulations. In legislation, in terms of legislation, local governments have different requirements for urban surveillance. According to San Francisco's Stop Secret Surveillance Ordinance (05/06/2019), the local government has made clear regulations on the surveillance of public areas. Ordinance amending the Administrative Code to require that City departments acquiring Surveillance Technology, or entering into agreements to receive information from non-City owned Surveillance Technology, submit a Board of Supervisors approved Surveillance Technology Policy Ordinance, based on a policy or policies developed by the Committee on Information Technology (COIT)... require each City department that owns and operates existing surveillance technology equipment or services to submit to the Board a proposed Surveillance Technology Policy Ordinance governing the use of the surveillance technology; and requiring the Controller, as City Services Auditor, to audit annually the use of surveillance technology equipment or services and the conformity of such use with an approved Surveillance Technology Policy Ordinance and provide an audit report to the Board of Supervisors [38]. In addition to San Francisco, Oakland has also made explicit requirements for urban surveillance. In Chapter 9.64 of Oakland, California—Code of Ordinances, Regulations on City's Acquisition and Use of Surveillance Technology requires that: (A) A description of how the surveillance technology was used, including the type and quantity of data gathered or analyzed by the technology, (B) Whether and how often data acquired using the surveillance technology was directly shared with outside entities, the name of any recipient entity, the type(s) of data disclosed, under what legal standard(s) the information was disclosed, and the justification for the disclosure(s), (C) Where applicable, a breakdown of what physical objects the surveillance technology hardware was installed upon; using general descriptive terms so as not to reveal the specific location of such hardware; for surveillance technology software, a breakdown of what data sources the surveillance technology was applied to; (D) Where applicable, a breakdown of where the surveillance technology was deployed geographically, by each police area in the relevant year, (E) A summary of community complaints or concerns about the surveillance technology, and an analysis of the technology's adopted use policy and whether it is adequate in protecting civil rights and civil liberties [39]. Apart from urban surveillance, there are also corresponding legislation on the privacy of biometric information. In Illinois, Biometric Information Privacy Act (BIPA) requires private entities that obtain biometric information or identifiers to first inform the subject in writing that their information is being collected and stored, inform the subject of the specific purpose and term for collection and storage, and secure a written release from the subject. BIPA also prohibits the disclosure of biometric information without the subject's consent, unless an exception is met. Private entities also cannot sell, lease, trade, or profit from a person's biometric information. Further, BIPA requires a private entity in possession of biometric identifiers and information to develop a publicly available written policy establishing a retention schedule and providing guidelines for the permanent destruction of the information [40].

Not only are there government mandatory requirements, but in order to restrict companies from collecting and using private data such as user faces in a reasonable and compliant manner, non-governmental organizations also have corresponding industry regulations. According to California Consumer Privacy Act (CCPA), CCPA gives consumers more control over the personal information that businesses collect about them and the CCPA regulations provide guidance on how to implement the law. This landmark law secures new privacy rights for California consumers, including: The right to know about the personal information a business collects about them and how it is used and shared; The right to delete personal information collected from them (with some exceptions); The right to opt-out of the sale of their personal information; and The right to non-discrimination for exercising their CCPA rights. Businesses are required to give consumers certain notices explaining their privacy practices. The CCPA applies to many businesses, including data brokers [41].

3.2 European Private Privacy Laws

For user privacy data, General Data Protection Regulation (GDPR) in Europe provides comprehensive and detailed specifications. In Art. 4 GDPR Definitions: 14. 'biometric data' means personal data resulting from specific technical processing relating to the physical, physiological or behavioral characteristics of a natural person, which allow or confirm the unique identification of that natural person, such as facial images or dactyloscopy data [42]. The GDPR provides a comprehensive definition of user privacy and biological data, giving users as much privacy protection as possible from the source. The scope of application of this regulation is extremely wide. Any organization that collects, transmits, retains or processes personal information in all member states of the European Union is bound by this regulation. In addition, some European countries have their own legislation to protect user privacy. For example, the Data Protection Act 2018 in the United Kingdom has made the following provisions on the protection of personal data:

(1) The GDPR, the applied GDPR and this Act protect individuals regarding the processing of personal data, by

    (a) requiring personal data to be processed lawfully and fairly, on the basis of the data subject's consent or another specified basis,

    (b) conferring rights on the data subject to obtain information about the processing of personal data and to require inaccurate personal data to be rectified, and

    (c) conferring functions on the Commissioner, giving the holder of that office responsibility for monitoring and enforcing their provisions.

(2) When carrying out functions under the GDPR, the applied GDPR and this Act, the Commissioner must have regard to the importance of securing an appropriate level of protection for personal data, taking account of the interests of data subjects, controllers and others and matters of general public interest.

Other European countries also have corresponding laws, such as Bundesdaten-schutzgesetz (BDSG) in Germany and La loi Informatique et Libertés in France. According to statistics, 12 of the 28 EU countries have officially updated their domestic laws to embed GDPR, and 8 countries have informed the European Commission that they will complete their legislative procedures as soon as possible in the near future. For those EU countries that have not incorporated GDPR into their domestic laws. The European Union has issued a warning to these countries, asking them to update their legal systems as soon as possible, otherwise they will be brought to the European Court of Justice. GDPR puts forward new and stricter requirements for obtaining and managing personal information. It gives individuals clear rights and has an important impact on high-tech companies' management of customer data through labor, process, and technology. Not only that, GDPR also has an impact on customer trust. Trust is the cornerstone of the digital economy. Meeting the data privacy and security requirements required by the GDPR is fundamental to main-taining consumer trust and protecting corporate brands. At the same time, violations will be strictly prohibited. The GDPR has greatly increased the compulsory and accountability of data protection, and the penalty amount for violations has been increased to 20 million euros or 4% of the company's global annual turnover (the higher value of the two). It was reported on May 28, 2018, that American companies such as Facebook and Google became the first defendants under the GDPR Act.

4 Proposal to Trustworthy

In the above section, we discussed ethical issues and user privacy data risks in FRT application scenarios such as demographic aspect, smart-city surveillance, face payment, and face to DNA. In fact, in today's human life, there are far more such problems. So how to use FRT well and how to build trustworthy AI has become a problem that requires great attention and needs to be solved urgently. Trustworthy AI itself is a quite comprehensive and complex research topic, encompassing many aspects. The next part of this article will analyze technically, how to achieve multi-dimensional trustworthy AI in the following 5 fields in Fig. 7: (1) Explainability, (2) privacy, (3) Nondiscrimination & Fairness, (4) Integrity, (5) environmental friendliness.

Fig. 7 5 dimensions of Trustworthy AI: (1) Explainability, (2) privacy, (3) Nondiscrimination & Fairness, (4) Integrity, (5) environmental friendliness

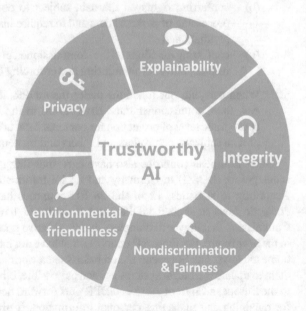

4.1 Explainability

First, we consider several common definitions of explainability of artificial intelligence. "The term 'explainable AI' or 'interpretable AI' refers to humans being able to easily comprehend through dynamically generated graphs or textual descriptions the path artificial intelligence technology took to make a decision." Keith Collins, executive vice president and CIO, SAS. And Stephen Blum, CTO and co-founder of PubNub said Explainable AI can be equated to 'showing your work' in a math problem. All AI decision-making processes, and machine learning don't take place in a black box—it's a transparent service, built with the ability to be dissected and understood by human practitioners. To add 'explanation' to the output, adding input/output mapping is key.

So why human need explainable AI? Regarding the level of trust in AI, in addition to the investigation of the original data, what is even more elusive is the seemingly mysterious and complicated machine learning models. If you cannot understand or accept these models, the results output by the models will inevitably be too questioned. Such problems are particularly obvious in some public issues related to social resources, such as employment, medical care, financial services, and law enforcement. There is such an example in the United States. Some district court judges have begun to use AI models to predict the sentence of suspects. We will not discuss the justice of this application for the time being. Assuming that it is legally and morally acceptable, it can be used by judges with confidence. A core consideration is whether they can understand the AI model and be able to think that the judgments made by AI are consistent with their logic.

In order to open the black box and make the algorithm model have a considerable Explainability, many scholars have proposed several representative and feasible solutions. Gebru, co-founder of Black in AI, in the research, her Google team and academics from the University of Toronto jointly proposed a method similar to Datasheets: Model Cards [43]. The model card is a record description attached to pages 1–2 of several released machine learning models, used to explain the multidimensional information of the model to the corresponding readers, in order to open the "black box" of the AI model. Figure 8 shows how many elements a model card can be divided into.

Then we will explain each element in the model card in detail.

A. Model details:

- Model developer (organization): the responsible natural person or team.
- Model release date: the date and time when the model was released and online.
- Model version: the version number of the model when it was released.
- Model type: traditional machine learning model, deep model, etc.
- Model references or reproduced papers, or open-source tools and sources for reference.
- Model reference specifications.

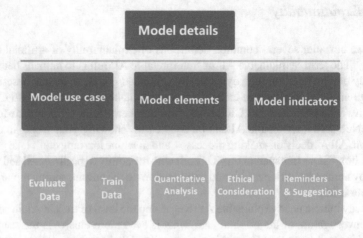

Fig. 8 The different elements included in Model Cards

- License to use the model.
- Feedback on the model (contact information).

B. Model use cases:

- User use cases or scenario descriptions of model applications, such as anti-fraud in credit card transactions in real time.
- The users (groups) for which the model is applied.
- Restrictions or prohibited use cases of model application, for example, which groups cannot be used for face recognition.

C. Model elements:

- Description of groups applicable to the model: especially those involving natural persons, clearly define applicable groups, as well as model performance and differences between different groups.
- Description of the external equipment or system to which the model is applicable: For example, in some CV scenarios, different camera models may cause differences in the performance of the model.
- Description of the external environment that the model is suitable for: Similarly, in the CV scene, different lighting conditions, ambient temperature and humidity may cause the model's effect to drift or even error.
- Other "silent" elements: In addition to the above elements, there are also elements that may be hidden under the surface that will affect the effect of the model.
- Evaluation elements: Will the design of some evaluation indicators affect

the comprehensiveness of the model investigation?

D.   Model indicators:

- What are the indicators selected by the model? What are the reasons for choosing these indicators?
- Indicator thresholds. What are the reasons for adjusting the default thresholds of some key indicators?
- What are the ways to deal with uncertainty and variability?

E.   Evaluate data:

- Description of the data set used for evaluation and the reason for choosing this part of the data set.
- Has this part of the data set been preprocessed? What is the way of preprocessing?

F.   Train data:

- You can refer to the contents of the Datasheets above for citation or condensed retelling so far.

G.   Quantitative analysis:

- Go back to the framework of the model elements and output the quantitative results of the analysis within different elements and print them into the model card.
- Between different model elements, output quantitative analysis results.

H.   Ethical considerations:

- During the development process, is private data used?
- Will the model be applied to matters that are critical to humans, such as safety, medical care? etc.
- Are there any risks and harm to some people in the use of the model?
- In the process of model development, whether some methods to reduce risks have been implemented.
- In the list of model use cases, are there any controversial use cases?

I.   Reminders & Suggestions:

- List the relevant information of the above models that are not included but are very important.

4.2 Privacy

Privacy is a critical issue for all types of data systems, but it is especially critical for AI since the sophisticated insights generated by AI systems often stem from data that is more detailed and personal. Trustworthy AI must comply with data regulations and only use data for the stated and agreed-upon purposes [44]. Laws and regulations such as GDPR and CCPA have made detailed regulations on how companies should collect, store, transmit, use, and destroy user privacy data. In the next section, we will also focus on the laws and regulations related to each country or region.

According to the Centre for Data Ethics and Innovation (CDEI) of the UK government, Privacy is a fundamental right. Organizations have an obligation to protect privacy, and must consider important legal, ethical, and reputational concerns when working with personal or sensitive data. Our report on public sector data sharing found that these concerns can lead to risk aversion that may inhibit data from being fully utilized to provide benefits for society. The use of privacy enhancing technologies (PETs) can help manage and mitigate some of the risks involved, potentially unlocking avenues to innovation [45].

In the broadest sense, a privacy enhancing technology is any technical method that protects the privacy of personal or sensitive information. This definition includes relatively simple technologies such as ad-blocking browser extensions, as well as the encryption infrastructure we rely on every day to secure the information we communicate over the internet. Of particular interest to the CDEI is a narrower set of emerging PETs. This is a group of relatively young technologies which are being implemented in an increasing number of real-world projects to help overcome privacy and security challenges. This set of emerging PETs is not rigidly defined, but there are a handful of technologies and techniques that are most prominent in current conversations. These include:

- Homomorphic encryption, which allows computations to be performed on encrypted data.
- Trusted execution environments, which can protect code and data in a processing environment that is isolated from a computer's main processor and memory.
- Secure multi-party computation, in which multiple organizations collaborate to perform joint analysis on their collective data, without any one organization having to reveal their raw data to any of the others involved.
- Federated analytics, an approach for applying data science techniques by moving code to the data, rather than the traditional approach of collecting data centrally.
- Differentially private algorithms, which enable useful population-level insights about a dataset to be derived, whilst limiting what can be learned about any individual in the dataset.
- Synthetic data, the generation of data that is statistically consistent with a real dataset. This generated data can replace or augment sensitive data used in data-driven applications.

4.3 Nondiscrimination & Fairness

'Bias in AI' has long been a critical area of research and concern in machine learning circles and has grown in awareness among general consumer audiences over the past couple of years as knowledge of AI has grown. It's a term that describes situations where ML-based data analytics systems show bias against certain groups of people. The previous part of our article also focused on specific bias issues using Demographic issue as an example. These biases usually reflect widespread societal biases about race, gender, biological sex, age, and culture [46]. There are two types of bias in AI. One is algorithmic AI bias or 'data bias', where algorithms are trained using biased data. The other kind of bias in AI is societal AI bias. That's where our assumptions and norms as a society cause us to have blind spots or certain expectations in our thinking. Societal bias significantly influences algorithmic AI bias, but we see things come full circle with the latter's growth. To build a trustworthy AI model, AI systems need to avoid unfair bias and provide for accessibility and universal design. Stakeholders who may be affected by an AI system should be considered and involved [47].

In the previous Demographic aspect section, we analyzed that expert who design algorithms may be biased in the algorithm model due to the manifestation of self-awareness; and because the data used may not be comprehensive enough, they can only reflect the objective reality of some groups, which will also cause Models trained using this data set will be biased against specific groups. Then we give some technical methods proposed by some teams and experts to alleviate or eliminate AI bias.

From the perspective of source data, experts of MOSTLY AI demonstrate that bias-corrected synthetic data can address both privacy and fairness concerns to allow for utilizing and democratizing big data assets while keeping the risks at a minimum. Their Synthetic Data Platform could generate highly accurate, statistically representative synthetic data at scale such as synthetic customer records along with purchase histories. The synthetic data, however, can be shared safely without privacy concerns since these artificial people do not really exist and the privacy of your actual customers, the real data subjects, remains protected. Synthetic data generation doesn't need to stop at privacy protection though. As they generate the data from scratch, they can model and shape it to fit different needs. A beautiful example of this is NVIDIA's styleGAN, where a conditional generation of synthetic images allows for adding smiles or sunglasses to faces or changing hair and skin color. Through the technology of bias-corrected synthetic data, on the one hand, people who really live in society do not need to worry about the leakage of their privacy, and on the other hand, scientific researchers engaged in AI development can obtain more comprehensive large amounts of data suitable for training models.

In addition, in the use of dataset, scientific researchers should be more rigorous and responsible. In Facial recognition from DNA using face-to-DNA classifiers [48], they present a complementary avenue for support in current DNA investigations and propose to match a probe DNA profile against a database of known facial profiles. In contrast to DNA phenotyping, the idea is not to predict facial characteristics from DNA, but instead to predict DNA aspects from 3D facial shape using face-to-DNA classifiers; hence, all information is estimated from existing 3D facial images in a database. In order to achieve their purpose, in data sampling process, their first study cohort (GLOBAL) consisted of n = 3295 unrelated and genetically heterogeneous individuals recruited from a variety of sites worldwide to identify the genomics background of a person in the context of genetic variation observed in diverse populations.

The author's sampling method meets the requirements of fairness, so that the model could not be affected by the one-sided data set as far as possible, and the output of the model will not discriminate against some groups.

4.4 Integrity

Self-driving cars give people great expectations for the future, but at the same time we should not let our guard down completely. Imagine this scenario. When an accident is about to happen to an autonomous vehicle during driving, whether it chooses to slow down and stay in the same lane to cause the owner to be injured, or to protect the owner and choose to change lanes but endangering the safety of others. This is a question that automobile companies need to think about. In addition, other industries are also facing the same problem. Under normal circumstances, human beings are aware of what they are thinking, but for smart machines, users may not know what the device is about to do, or for what purpose it has done a certain behavior.

These concerns of users show their distrust of intelligent machines, which is also a problem that AI product designers need to solve: how to allow users to build trust in artificial intelligence technology, and how to make the autonomous operation of intelligent machines conform to the human social value system. As we all know, the social value system has many aspects and covers too much, so a trustworthy smart device should also have integrity in different aspects. For humans to trust an AI, the algorithms, software and production deployment systems that train AI models and deliver AI predictions must behave within specified norms and in a predictable and understandable way—i.e., with ML Integrity. ML Integrity is the core criterion that a machine learning (or deep learning, reinforcement learning etc.) algorithm must demonstrate in practice and production to drive trust [49]. Machine learning integrity is a necessary condition for developing trustworthy AI systems. Machine learning integrity can help ensure that AI systems generate the output according to a developer's predefined operational and technical parameters. With machine learning integrity, developers can make sure that AI systems work as they are intended to. Also, developers can set up certain limitations for AI systems that can be used to regulate the usage of AI. In this manner, developers can design trustworthy AI systems that produce accurate results by following predefined conditions [50]. Other domains, including other software arenas, have established the concept of integrity as a core element for trust.

The author of ML Integrity: Four Production Pillars for Trustworthy AI states that production ML systems have many moving parts. At the center is the Model, the trained AI algorithm that is used to generate predictions. However, models must be trained and then deployed in production. Many factors (example: the training dataset, training hyper-parameters, production code implementation, and incoming live production data) combine to generate a prediction. For this entire system and flow to behave with integrity, four pillars must be established. These are shown in the Fig. 9.

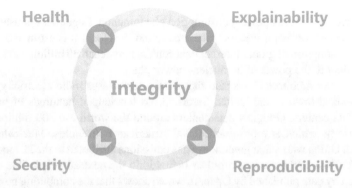

Fig. 9 Four parts contained in an AI system to achieve integrity

- ML Health: the ML model and production deployment system must be healthy—i.e., behaving in production as expected and within norms specified by the data scientist.
- ML Explainability: it must be possible to determine why the ML algorithm behaved the way that it did for any prediction and what factors led to the prediction.
- ML Security—the ML algorithm must be healthy and explainable in the face of malicious or non-malicious attacks—i.e., efforts to change or manipulate its behavior.
- ML Reproducibility: All predictions must be reproducible. If an outcome cannot be faithfully reproduced, there is no way to know for sure what led to the outcome or debug issues.

4.5 Environmental Friendliness

Climate change is a pressing issue for people worldwide, but artificial intelligence could be the key to helping save our environment. AI is applying the power of machine learning to finding patterns in data that can help spot trends. At the same time, AI's energy consumption is high. Data centers are critical for storing the large amounts of data needed to power AI systems but demand a huge amount of energy. In addition, training advanced artificial intelligence systems, including deep learning models, can require high-powered GPUs to run for days at a time [51].

Globally, data centers were estimated to use between 196 terawatt hours (TWh) [52] and 400 TWh [53] in 2020. This would mean data centers consume between 1 and 2% of global electricity demand. In order to allow Internet users around the world to enjoy the convenience of the Internet 24 h a day, major Internet companies have been providing round-the-clock uninterrupted services. However, due to the long-term operation of the servers of major websites, the power consumption is astonishing. According to a survey conducted by reporters from the New York Times for more than a year, many Internet companies and their data centers in the United

States have an alarming power consumption phenomenon. Regardless of user needs, the data centers of major websites operate around the clock. It is estimated that the power consumption of global Internet data centers may reach 30 billion watts, which is equivalent to the power of 30 nuclear power plants.

Data center architect Gross said that a data center uses more electricity than a medium-sized town in the United States. Gross has helped hundreds of websites design data centers. Google's data centers around the world use 300 million watts of electricity, which is more than 30,000 American households. Meanwhile, the European Union warns that greenhouse gas emissions attributed to the IT industry—currently 2%—could rise sevenfold to 14% within the next 20 years. This trend is supported by data published by OpenAI, which shows that the computing power that has been required for key AI landmarks over recent years, such as DeepMind's Go-playing programme AlphaZero, has doubled roughly every 3.4 months—a 300 k-fold increase between 2012 and 2018 [54].

Take the neural network that is now widely used in AI as an example. The flexibility of the neural network comes from (researchers) feeding many inputs to the model, and then the network combines them in a variety of ways. This means that the output of the neural network comes from the application of complex formulas, not simple formulas. In other words, the amount of calculation of the neural network will be very large, and the computing power requirements of the computer will be extremely high. For example, when Noisy Student (an image recognition system) converts the pixel value of an image into the probability of an object in the image, it is implemented by a neural network with 480 million parameters. The training to determine the value of such a large-scale parameter is even more jaw-dropping: because this training process only uses 1.2 million labeled images.

On the one hand, deep learning has strong flexibility; but on the other hand, this flexibility is based on huge computational costs. As shown in Fig. 10, according to existing research, by 2025, the error level of the best deep learning system designed to identify target objects in the ImageNet dataset should be reduced to only 5%.

However, the computing resources and energy consumption required to train such a system are huge, and the carbon dioxide emitted is about as much as the carbon dioxide produced by New York City in a month (Fig. 11).

There are two main reasons for the increase in computational cost: (1) To improve performance through factor k, at least k to the second power or even more data points are required to train the model; (2) excessive parameterization. Once the phenomenon of over-parameterization is considered, the total computational cost of the improved model is at least k to the 4th power. The small "4" in this index is very expensive: an improvement of 10 times requires at least an increase of 10,000 times the amount of calculation.

In response to this series of problems, researchers at UC Berkeley and Google have written about how this can be done. There are three things that can be controlled: the design of your algorithms, the hardware on which you train them, and the availability of green electricity to the data centers in which they run. Large companies such as Google may have the luxury of being able to control all three of these key features, but your average AI-adopting company probably does not. It is therefore imperative

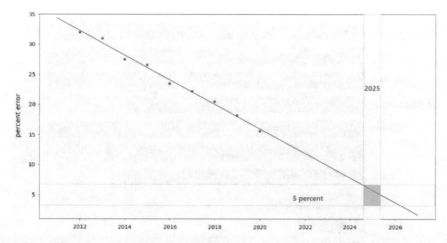

Fig. 10 The error percent level of the best deep learning system designed to identify target objects in the ImageNet dataset

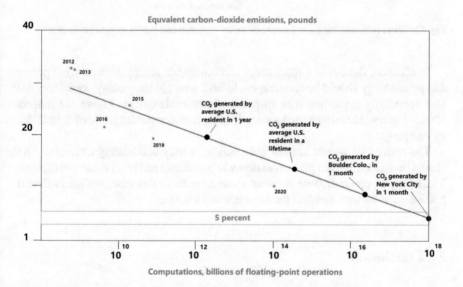

Fig. 11 The emitted carbon dioxide level with the increase of computation

that those developing the AI make sure they consider sustainability throughout the design and training process.

AI can also be used in green projects, as many companies already do. This ranges from the use of AI in manufacturing to reduce waste and energy use, to companies like UK startup Space Intelligence, who are applying machine learning and AI to satellite data to address environmental concerns such as reforestation and biodiversity conservation so that industries can take corrective steps.

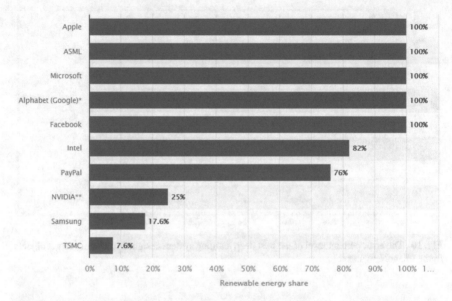

Fig. 12 Share of renewable energy used among leading tech companies worldwide in 2020

In addition, the use of clean energy and renewable energy such as wind power and solar energy should be encouraged. In fact, now big technology companies are also voluntarily requesting their own use of renewable energy. Figure 12 reflects Share of renewable energy used among leading tech companies worldwide in 2020, by company:

The above five aspects are not independent, a truly outstanding trustworthy AI should be able to give full play to creativity to coordinate and unify these five aspects. There may be contradictions between some aspects. In this case, designers should make a good balance and find the most optimal solution.

5   Conclusion

In this article, we first start with several representative practical application scenarios of FRT, introduce the role of FRT in various industries, and analyze some ethical issues in the current use of FRT from multiple aspects. After that, we have detailed examples of some regulations that may be violated in the process of using FRT in the United States and Europe from GDPA, CCPA and other laws and regulations. Finally, in order to establish trustworthy AI, we start from a technical point of view and give some representative technical methods from the five aspects of ABCDE. However, due to the complexity of the social value system and the continuous changes in human thinking and concepts over time, the construction of trustworthy AI will encounter many problems in the future and needs to keep pace with the times.

References

1. Roussi A (2020) Resisting the rise of facial recognition. Nature 587:350–353
2. Noorden RV (2020) The ethical questions that haunt facial-recognition research. Nature 587:354–358
3. Castelvecchi D (2020) Is facial recognition too biased to be let loose? Nature 587:347–349
4. Kelion L (2019) Crime prediction software 'adopted by 14 UK police forces. BBC, 4 Feb 2019 (Tech)
5. Rauenzahn B, Chung J, Kaufman A (2021) Facing bias in facial recognition technology. Rights, Saturday Seminar, Penn program on regulation
6. Nations United (2020) Bias, racism and lies: facing up to the unwanted consequences of AI. UN News, 30 Dec 2020
7. BBC (2019) Artificial intelligence: algorithms face scrutiny over potential bias. BBC, 20 March 2019 (Tech)
8. Gebru T (2020) Race and gender. In: The Oxford handbook of ethics of AI
9. Manyika J, Silberg J, Presten AB (2019) What do we do about the biases in AI? Harvard Bus Rev
10. Brey P (2004) Ethical aspects of facial recognition systems in public places. J Inf Commun Ethics Soc 2
11. Bu Q (2021) The global governance on automated facial recognition (AFR): ethical and legal opportunities and privacy challenges. Int Cybersecur Law Rev 2:113–145
12. Benniao Technology (2020) Google caught off guard! The privacy of millions of users has been "leaked", demanding $5 billion in compensation. 27 Oct 2021. https://baijiahao.baidu.com/s?id=1668544543962194461&wfr=spider&for=pc
13. Clever S, Crago T, Polka A et al (2018) Ethical analyses of smart city applications. Urban Sci 2
14. Stein MI (2020) New Orleans City Council bans facial recognition, predictive policing and other surveillance tech. The LENS, 18 Dec 2020
15. McDonough A (2020) The NYPD has a surveillance problem. City&State, 15 Oct 2020
16. Kitchin R (2019) The ethics of smart cities. RTE, 27 April 2019
17. Wei M, Rodrigo V (2021) The impact of face recognition payment in the economic. In: DESD 2021. Atlantic Press
18. Che SP, Nan D, Kamphuis P, Kim JH (2021) A comparative analysis of attention to facial recognition payment between China and South Korea: a news analysis using Latent Dirichlet allocation. In: HCI International 2021, International conference on human-computer interaction, pp 75–82
19. Kawakami T, Hinata Y (2019) Pay with your face: 100 m Chinese switch from smartphones. NIKKEI Asia, 26 Oct 2019
20. Selfin M (2020) Is the rest of the world ready for facial pay? Payments J, 10 March 2020
21. Dan Z (2021) Apps barred from indiscriminate collection of unnecessary personal information. Global Times, 28 July 2021
22. Allen K (2019) China facial recognition: law professor sues wildlife park. BBC, 8 Nov 2019
23. Dean S (2020) Forget credit cards—now you can pay with your face. Creepy or cool? Los Angeles Times, 14 Aug 2020
24. Liu Y, Yan W, Hu B (2021) Resistance to facial recognition payment in China: the influence of privacy-related factors. Telecommun Policy 45:102155
25. Apple Inc. (2021) About Face ID advanced technology. 29 Oct 2021. https://support.apple.com/en-us/HT208108
26. THALES. Facial recognition issues. https://www.thalesgroup.com/en/markets/digital-identity-and-security/government/inspired/facial-recognition-issues
27. Analytic IT solutions. 7 simple tips for optimizing a website's user experience. https://www.analytixit.com/2020/06/17/7-simple-tips-for-optimizing-a-websites-user-experience/
28. Frederickx I (2019) From face to DNA: new method aims to improve match between DNA sample and face database. KU LEUVEN, 11 June 2019

29. White JD, Indencleef K, Naqvi S et al (2020) Insights into the genetic architecture of the human face. Nat Genet 53:45–53
30. Weinberg M, Shaffer JR (2020) Researchers scan DNA to learn how facial features form. Pittwire
31. Hereward CCAJ (2018) How accurately can scientists reconstruct a person's face from DNA? Smithsonian
32. Marano LA, Fridman C (2019) DNA phenotyping: current application in forensic science. Res Reo Forensic Med Sci 9:1–8
33. ACOG Committee Opinion (2008) Ethical issues in genetic testing. ACOG (410)
34. Goh K, Cusick ME, Valle D et al (2007) The human disease network. Proc Natl Acad Sci 104:8685–8690
35. CODEX genetics (2020) Codex is committed to keep personal data and genetic information confidential. https://www.codexgenetics.com/blog/en/Protecting-the-Privacy-of-Genetic-Information/
36. Richardson R (2021) Facial recognition in the public sector: the policy landscape. GMFUS
37. Almeida D, Shmarko K, Lomas E (2021) The ethics of facial recognition technologies, surveillance, and accountability in an age of artificial intelligence: a comparative analysis of US, EU, and UK regulatory frameworks. AI Ethics
38. Administrative Code—Acquisition of Surveillance Technology. Stop Secret Surveillance Ordinance. Electronic Frontier Foundation (05/06/2019)
39. Okland. Chapter 9.64 - Regulations on City's Acquisition and Use of Surveillance Technology. Oakland, California—Code of Ordinances
40. Polsinelli PC (2021) Past, present and future: what's happening with Illinois' and other biometric privacy laws. The National Law Review XI(308)
41. California Consumer Privacy Act (CCPA). https://oag.ca.gov/privacy/ccpa
42. General Data Protection Regulation (GDPR). https://gdpr-info.eu/
43. Mitchell M, Wu S, Zaldivar A et al (2018) Model Cards for Model Reporting. arXiv:1810.03993[cs.LG]
44. Saif I, Ammanath B (2020) 'Trustworthy AI' is a framework to help manage unique risk. MIT Technology Review
45. GOV.UK, Privacy enhancing technologies for trustworthy use of data. https://cdei.blog.gov.uk/2021/02/09/privacy-enhancing-technologies-for-trustworthy-use-of-data/
46. LEXALYTICS (2021) Bias in AI and machine learning: sources and solutions. 30 Oct 2021. https://www.lexalytics.com/lexablog/bias-in-ai-machine-learning
47. Hickman E, Petrin M (2021) Trustworthy AI and corporate governance: the EU's ethics guidelines for trustworthy artificial intelligence from a company law perspective. Springer
48. Sero D, Zaidi A, Li J et al (2019) Facial recognition from DNA using face-to-DNA classifiers. Nature, 2557
49. Talagala N (2019) ML Integrity: four production pillars for trustworthy AI. Forbes, 29 Jan 2019
50. Joshi N (2019) How we can build trustworthy AI. Forbes, 30 July 2019
51. Labbe M, Schmelzer R (2021) AI and climate change: the mixed impact of machine learning. TechTarget,
52. Masanet E, Shehabi A, Lei N, Smith S, Koomey J (2020) Recalibrating global data center energy-use estimates. Science 367:984–986
53. Hintemann R (2020) Data centers 2018. Efficiency gains are not enough: Data center energy consumption continues to rise significantly—cloud computing boosts growth
54. Mullins B (2021) Time to tackle AI's impact on the environment. Sifted

A Technical Review on Driverless Vehicle Technologies in Smart Cities

Yijie Zhu, Richard Jiang, and Qiang Ni

Abstract In the past decade, due to the progress of artificial intelligence, the development of driverless vehicle technologies is changing with each passing day. Because of this, driverless vehicles have been introduced into smart city and become one of the most popular urbanization solutions. It is expected to reduce the traffic burden, improve land use efficiency and expand the scale of cities. This paper investigates the main areas of driverless vehicles, such as sensors and hardware, localization and mapping, perception, planning and decision making, etc. This paper also introduces the database for driverless vehicles and the application of driverless vehicles in smart city.

1 Introduction

The definition of driverless vehicle in UCSUSA [1] is: "Self-driving vehicles as cars or trucks in which human drivers are never required to take control to safely operate the vehicle." Driverless vehicle is also known as autonomous vehicle. This kind of vehicles can perceive the surrounding environment through sensors and control, navigate and drive vehicles through the automated driving system. In the past decade, driverless vehicles have become popular due to the need to prevent accidents, reduce carbon emissions and relieve urban traffic pressure. The integration of technology and cities is viewed as a potential solution for overcoming the difficulties of urbanization, for example, environmental change, gridlock and greenhouse gas (GHG) emissions [2]. As an indispensable piece of the city, transportation accounts for about one fourth to one third of greenhouse gas emissions. Innovation for the purpose of intelligent city transportation is becoming a key concept in the contemporary urban policy plan to address the adverse impacts of transportation. The concept of smart city transportation is characterized by integrating sustainable and intelligent vehicle technology and cooperative intelligent transportation system through cloud server

Y. Zhu (✉) · R. Jiang · Q. Ni
LIRA Center, Lancaster University, Lancaster L1 4YW, UK
e-mail: y.zhu43@lancaster.ac.uk

© Springer Nature Switzerland AG 2022
R. Jiang et al. (eds.), *Big Data Privacy and Security in Smart Cities*,
Advanced Sciences and Technologies for Security Applications,
https://doi.org/10.1007/978-3-031-04424-3_3

and vehicle network based on big data [3]. As a part of the smart city transportation system, one of the most advanced applications is driverless vehicles.

With the rapid development of electronics, information and communication technology, especially in the field of computer vision, driverless vehicles have also made some achievements. This has opened up a lot of research and development fields. Although they are ultimately connected with driverless vehicles, they are corresponding to very different fields. Because of this, it is necessary to investigate the related technologies and the urgent challenges of driverless vehicles.

The first major autonomous driving competition was DARPA Grand Challenge organized by the US Department of defense in 2004 [4]. The goal of this challenge is for driverless vehicles to drive 150 miles automatically off-road without any human intervention. Although no one completed the challenge at that time, many similar challenges have emerged since then. Autonomous driving in urban scenes is the most popular and difficult task in this field. DARPA Urban Challenge held in 2007 [5]. In the end, six teams completed the challenge. Although there is a gap between the urban scene in this challenge and the actual scene, it stimulates the enthusiasm of researchers in this field. Since then, many autonomous driving challenges for urban scenes have been held in various countries. After years of research and practice, driverless vehicle architecture is gradually established. Due to the complexity of the automated driving task, most of the automated driving systems of driverless vehicles are divided into several modules, each of which is responsible for a specific task. End to end autonomous driving is a trend developed recently, especially the application of deep learning in driverless vehicles has greatly increased, which promotes this trend.

The rest of this paper is divided into six parts. The second section outlines the current opportunities and challenges, the third section introduces the architecture of driverless vehicles, the fourth section introduces sensors and hardware, the fourth section introduces localization and mapping, the fifth section introduces perception related tasks and technologies in detail, the fifth section introduces planning and decision making, and the sixth section introduces V2X under the background of smart city. Section seventh introduces the relevant datasets of autonomous driving.

2 Prospects and Challenges

At present, a widely recognized taxonomy is proposed by Society of Automotive Engineers (SAE) [6]. In this taxonomy, there are 6 levels of autonomous driving. Level zero is no driver automation. Even with the assistance of the active safety system, the driver still performs the driving task. Level one is driver assistance. The system offers driving support for single operation of the steering wheel controlling and deceleration through surrounding situation, and other driving tasks are finished by drivers. Level two is partial driving automation. The system offers driving support for multiple operations through surrounding situation, and other driving activities are finished by drivers. Level three is conditional driving automation. Conditional driving automation refers to autopilot in some specific scenarios (highway, overpass, road congestion, etc.), the drivers still need to monitor driving activities. Level four is high driving automation. The vehicle is driven by the system. According to the system requirements, the driver doesn't need to react to all the requests. Level five is full driving automation. This level of system allows all occupants in the vehicle to engage in other activities without monitoring.

At present, most of the cars on the market belong to level one. Many auto manufacturers devoted to the development of autonomous driving have launched products reaching level two. Level two cars can achieve adaptive cruise control (ACC) and lane keeping assist (LKA) at the same time. Although the driver still needs to concentrate on driving, the automatic control of the system liberates the driver's hands and feet, so that the driver can focus on the road conditions and reduce the driver's distraction and fatigue. According to the taxonomy of autonomous driving, it can be found that autonomous driving can reduce the operating burden of human drivers, and thus reduce the traffic accidents caused by human driver's wrong behaviors (distraction, fatigue, over speed, etc.). Considering the aging problem, the aging population is gradually expanding, autonomous driving can solve their traffic problems and improve the quality of life. In addition, in the autonomous driving task based on the background of smart city, using the Internet of vehicles to reasonably allocate the traffic flow of vehicles in the region can solve the problem of traffic congestion. In the context of autonomous driving and smart city, Mobility as a Service (MaaS) can completely comprehend and share the resources given by the entire city transportation, to accomplish consistent, safe, and convenient travel service. In connection with the role of transportation in promoting social development, autonomous driving has a good prospect.

There are still some challenges in the development of autonomous driving. At present, only Tesla and Audi claim to have reached level three. Level 3 vehicles can completely take over the driving task in a specific scene until the system notifies the driver to take over again. From literature [7], when the system completely takes over the vehicle driving, the risk of collision accidents is increased. In view of the public's concern about driverless vehicle accidents, autonomous driving technology needs to be improved. At present, it is still a challenge to reach level 3 and ensure the reliability of autonomous vehicles. In addition to the self-driving vehicle itself, the security and reliability of IoV is also a challenge in the smart city scenario. This kind of technology using the Internet has the risk of being attacked by hackers. If the attack is successful, the impact will be very negative.

3 Architecture of Automated Driving Systems

The autonomous driving system can be classified from two angles: connectivity and algorithm. From the perspective of connectivity, the automated driving system can be divided into independent systems and V2X systems. From the perspective of

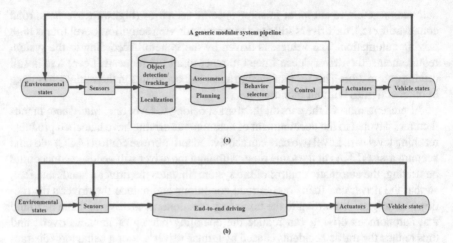

Fig. 1 Information flow diagrams of: **a** a generic modular system, and **b** an end-to-end driving system

algorithm, the systems can be divided into modular systems and end-to-end systems (Fig. 1).

Independent systems mainly means that autonomous driving depends on the intelligence of the vehicle, and does not need other vehicles or external sensors. Baidu Apollo 3.0 is a kind of independent systems [8]. Connected system refers to the automated driving system which is composed of V2X devices to form the IoV for distributed processing. V2X means "vehicle to everything", that is the information exchange between vehicles and the external world [9]. The system using V2X can sense a series of traffic information such as real-time road condition, road information, pedestrian information through information exchange with external base stations or other vehicles. Depending on this information, the system can do many operations such as avoiding congestion and avoiding pedestrians. Baidu Apollo 5.0 is a kind of connected systems [8]. At present, most of the automated driving systems belong to independent system. Compared with connected systems, independent system is low cost and easy to develop. At present, the automated driving system is not yet mature and still needs a lot of development and testing. It is reasonable to choose independent systems. The automated driving system based on V2X is still in the stage of research and exploration. In a city with thousands of vehicles, the application of this system is still a challenge. The difficulty is mainly caused by the network. How to allocate IP, route and ensure the security are the problems to be solved. The V2X based automated driving system conforms to the concept of smart city especially the network facilities and has a good development prospect.

Modular systems refer to a pipeline structure composed of several components or modules responsible for specific functions. The automated driving system is mainly composed of Localization and Mapping, Perception and Decision Making. A typical pipeline [8] receives external information from sensors, delivers it to localization module, passes through each module in sequence, and finally updates the status of the vehicle. For the end-to-end system, the system will directly output the instructions to the vehicle (speed increase and decrease, steering wheel rotation) according to the input information [10, 11]. The advantage of modular systems is that it can develop different components separately, which makes it easier to solve a single functional problem. The combination and coverage of various modules also reduce the difficulty of development, and the reliability of the system is high. The end-to-end autopilot system is very simple and very close to human driving habits, but it lacks interpretability. Although it has a good prospect and has achieved good results, it will take time for the system to be used safely in the urban environment.

4 Sensors and Hardware

At present, the automated driving system is widely applied to the selection and application of sensors and related hardware. As a technology involving traffic safety, automated driving system has higher requirements for improving perception and robustness. It is a common practice to use a variety of sensors to collect external information in an automated driving system. Sensors are mainly used to sense external information, including vehicles, pedestrians, buildings and so on, to provide the system for the next operation. The most common sensors are camera, radar, lidar and ultrasonic (Fig. 2).

Fig. 2 Sensors and hardware overview of Baidu Apollo [8]

Camera is a widely used sensor. The advantage of the camera is that it can accept the color information of the external environment. Color information plays a key role in detecting the vehicles, pedestrians, especially the signal lights. With the rapid development of computer vision technology, the object detection algorithm based on computer vision has made many achievements, and the method of computer vision has also been introduced into the field of automated driving, which has made good progress. Compared with radar, lidar and ultrasonic, the camera is passive in receiving image information and will not actively release signals, so there is no risk of interfering with other systems. The disadvantages of the camera are also obvious. The illumination condition of the camera has a great influence on the information collection. The poor illumination condition can easily affect the quality of the information collection and make the system make a wrong decision. In addition, monocular camera is difficult to obtain depth. Although some studies have improved this problem, compared with active sensors such as radar, the gap of monocular camera is still large.

Radar is an active sensor. Radar emits radio waves, bounces back from the object, and measures the time of each bounce to get the depth of the environment. Radar is a mature technology, which has many applications in automated driving vehicles. Although active sensors may interfere with other components, this problem can be avoided by reasonable design. Lidar is a popular sensor in autonomous driving. Compared with radar, lidar uses infrared instead of radio waves to detect. Under 200 m, the accuracy and response speed of lidar are better than radar, and the interference to other parts of the system is relatively small. However, infrared is more vulnerable to rain, fog, snow and other weather effects, resulting in reduced accuracy. In addition, the size of lidar is larger than that of radar.

The working principle of ultrasonic is to measure the distance by the time difference between sending ultrasonic wave through the ultrasonic transmitter and receiving ultrasonic wave through the receiver. The ultrasonic will not be affected by water and dust. Even if there is a small amount of sand, the detection effect of ultrasonic will not be affected. However, the effective distance of ultrasonic is limited, generally less than three meters, which limits the application of ultrasonic.

Global Positioning System (GPS) is an indispensable technology for driving localization. The GPS comprises of 32 GPS satellites in space, one primary control station, three data injection stations, five observing stations and a GPS receiver as the client. No less than three of the satellites are expected to speedily determine the location and elevation of the client on the earth. Even in terrible climate, GPS can in any case keep signal. Inertial Measurement Unit (IMU) is a sensor to detect acceleration and rotation. After processing the IMU data, the system can get the displacement and rotation information of the vehicle in real time. GPS and IMU are often used in automated driving system for localization.

5 Localization and Mapping

The purpose of localization is to make the automated driving system find its exact location. In the route planning task, the automated driving system needs accurate location of the vehicle's current position before it can make efficient planning. At present, there are three state of art Localization methods: GPS-IMU fusion, priori map and simultaneous localization and mapping (SLAM).

Many navigation devices are based on GPS. However, driverless vehicles are driven in a complex dynamic environment, especially in big cities. For GPS, it is easy to deviate due to the reflection of signal from complex environment. For a car with limited width and high speed, such deviation is likely to lead to traffic accidents. Although IMU error will increase with time, it can provide stable real-time location update in a short time. GPS-IMU fusion, which combines the two methods, can get real-time and accurate localization. Even so, in the complex urban environment, GPS-IMU fusion system still has limitations, especially when vehicles pass through tunnels and bridges, GPS is easy to stop working. GPS-IMU fusion system can only be used for high-level route planning.

The location method based on prior map is to make high precision map in advance. The automated driving system compares the external environment information with the prior map, and finds the highest matching place on the prior map to locate it. Compared with GPS-IMU fusion method, this method has better effect in complex urban environment, but its disadvantages are also obvious. The biggest problem of this method is that it costs too much to make high-precision maps. Secondly, the speed of map making may not catch up with the speed of environmental changes [12]. Considering the impact of seasonal changes on the environment, the cost of making a priori map will be quite high. Compared with the GPS-IMU fusion method, the instantaneity of this method is not enough.

Simultaneous localization and mapping (SLAM) is a kind of localization, navigation and mapping technology that was originally used for robot in unknown environment. The significance of slam for driverless vehicles lies in how to help vehicles perceive the surrounding environment and better complete the advanced tasks such as navigation, obstacle avoidance and path planning. Salm can complement the defects of GPS-IMU fusion and prior map in accuracy and timeliness. Considering the problems of complex environment and untimely map updating in practice, slam can be used as a good complementary method combined with the above two methods. With the improvement of deep learning, slam has a good performance, and will have a better performance in the field of automatic driving in the future [13].

6 Perception

Perceiving the external environment of vehicle and extracting the information required by the automated driving system is a key task of autopilot. Perception

Fig. 3 Example of 2D object detection

tasks involve many specific tasks, including 2D object detection (Fig. 3), semantic segmentation, moving object tracking, 3D object detection, road and lane detection.

The task of object detection is to extract the location, size and features of the object of interest. For autopilot systems, targets include, but are not limited to, vehicles, signal lights and pedestrians. 2D object detection generally refers to extracting the required information from the image, deciding whether there are specific types of objects in the image that the system needs to pay attention to, and determining their size through the bounding box. In 2D object detection tasks, state of the art methods is based on deep learning, which are mainly divided into two kinds. One is a single-stage detection, which uses a single network framework to generate object detection location and category prediction at the same time. The other is a two-stage detection framework, which first proposes the general regions of interest, and then classifies them through a separate classifier network. "You Only Look Once" (YOLO) [14] and "Single Shot multibox Detector" (SSD) [15] are two common single-stage detection frameworks. YOLO uses convolutional neural network (CNN) to extract features on a grid, and improves the speed of operation by reducing the resolution. Then a fully-connected neural network is used to predict the class probability and boundary box parameters of each grid cell and class. Yolo's disadvantage is that each grid only predicts one object, which is easy to cause missed detection, while SSD is improved. YOLO network has been continuously updated to YOLOv5 [16], which has a huge improvement in training speed and inference time, so it is very suitable for vehicle platform. Compared with the single-stage detection framework, the performance of the two-stage detection framework is better. Region proposal networks (RPN) is superior to several single-stage detection frameworks in target recognition and positioning accuracy. The limitation of the two-stage detection framework is that it requires high computing power and usually it is not easy to implement. For driverless system, single-stage detection algorithm often has fast inference time and low storage cost, which can meet the needs of real-time computing. Faster R-CNN [17] is a typical network using RPN. If the computing power can be further improved, RPN may replace the single machine detection network and become the mainstream detection algorithm of autonomous driving.

Semantic segmentation is a common method in driving scene understanding task. Some objects of interest are poorly marked with 2D bounding boxes, such as roads, traffic lines and sidewalks. For this type of object, semantic segmentation classifies every pixel, so it can play a good effect. At present, the relatively new methods are CenterMask [18] for instance segmentation and EfficientNet-based EfficientPS [19]. Dong et al. [20] proposes a real-time and high-performance semantic segmentation method of city street scene based on DCNN, which achieves a good tradeoff between accuracy and speed. This method has good real-time performance in the semantic segmentation of urban street scene. At present, the network used for semantic segmentation is still too slow and expensive to compute, but considering the advantages of semantic segmentation, it is expected that a network suitable for automated driving system will appear in the future.

At present, most of the environment perception algorithms are based on camera. The reason for this phenomenon is that the cost of the camera is low, the versatility is high, and the computer vision algorithm has been widely studied. However, the image that the camera can process is projection image, and the perception of scene scale is missing, which is easy to cause errors. Therefore, it is necessary to introduce 3D object detection into the target detection task. Lidar is a reliable way of 3D detection. The process of 3D object recognition is filtering, clustering and feature extraction. Lidar obtains point cloud data from the surrounding, and the detection methods can be divided into three-dimensional volume-based method and projection-based method (mapping 3D data to 2D plane). PointNet [21] is an efficient network, which can learn local and global features of point cloud data without pre-processing. In Voxel-FPN [22], a one-stage object detector, a novel one-stage 3D object detector is proposed to utilize raw data from lidar. MLCVNet [23] introduces self-attention mechanism into 3D object detection. The accuracy of volume-based method is generally higher than that of projection-based method, but the calculation is large. Although the accuracy of projection-based method is relatively low, it is more efficient.

Considering the extra cost of lidar, some researches try to use camera-based methods for 3D object detection. Because it is difficult for camera to obtain depth, camera-based methods usually use prior knowledge in geometry reasoning for 3D object detection. M3D-RPN [24] utilizes the geometric relationship between 2 and 3D perspectives, allowing 3D boxes to utilize convolution features generated in image space. Mono3D++ [25] takes the image as the input and generates the trajectory of the detected object in the world coordinate system. Combined with the vehicle shape and posture, it optimizes the projection between the generated 3D hypothesis and its 2D pseudo measurement. Shift R-CNN [26] uses a faster R-CNN network to regress the initial two-dimensional and three-dimensional object properties, and uses the projection matrix to combine it with the two-dimensional to three-dimensional geometric mapping problem.

It is important to estimate the course and speed of the dynamic target in the driving process of the driverless vehicle, so as to apply the motion model to track the object and predict the trajectory. In order to track and predict the target trajectory, a single sensor is not enough. Multiple camera systems, lidar and radar sensors are needed to obtain range information. To adapt to the impediments and vulnerabilities of various detecting modes, simultaneous interpreting is usually done by sensor fusion strategy. Normally utilized object trackers depend on simple information affiliation techniques followed by conventional filtering techniques. When tracking targets in high frame rate 3D space, the nearest neighbor method is usually sufficient to establish the association between targets. Point-cloud-based methods may also use similarity metrics. When tracking targets in high frame rate 3D space, the nearest neighbor method is usually sufficient to establish the association between targets. In the filtering step, for each object, the position is estimated by taking the mathematical mean of the object. Position estimation is normally refreshed by Kalman filter or particle filter. After filtering, the feature extracted from the image and 3D points is used to track the target based on segment matching. The target tracking method based on deep learning has also achieved good results, which is close to real-time detection. SiamRPN++ [27] is based on SiamFC, introduces RPN and avoids multi-scale test to improve performance and accelerate the speed. PrDIMP [28] predicts the conditional probability density of the target state to track the target, and greatly improves the tracking effect. [29] uses tensor and convolution LSTM to predict the spatiotemporal trajectory sequence. Tensor and convolution can better understand the spatiotemporal interaction between pedestrian and environment.

Boundary box methods are good for detecting objects, but not for continuous surfaces such as roads. Road identification is one of the important tasks of the automated driving system. In the above context, semantic segmentation is used to determine the feasible road surface, but in addition to extracting the semantics, the automobile driving system still needs to understand the road semantics in order to properly cross the road. It is challenging to understand the driveway and cross the intersection by relying solely on the driverless vehicle itself. Road and lane detection has a variety of methods. The easiest method is to determine the feasible driving area from the vehicle's perspective, then divide the road into lanes and determine the main lane of the vehicle. However, to achieve a Level3 and above driverless car, the automated driving system needs to be able to determine other lanes to make turns. This requires that the automated driving system understand the semantic information of the load. Deep learning is widely used in road detection tasks. In literature [30], a deep learning method is developed, which combines LIDAR point cloud and camera image for road detection. In smart city, V2X technology is supposed to be used in road identification, but it is still difficult to achieve.

7   Planning and Decision Making

Planning can be divided into two sub-tasks: route planning and path planning. The task of route planning is to find the route from the starting point to the final target on the road network. Path planning is usually regarded as the issue of finding the shortest path in a directed graph. The most classic path planning algorithm is A* algorithm, which has been used in various fields so far. A* algorithm is a goal directed algorithm. Arc flags [31] algorithm is also a goal directed algorithm, although this method takes a long time to preprocess, but the query speed is fast. There is also a separator-based method, by deleting a subset of vertices or arcs and calculating the overlay on it. This method can make the query faster. Customizable Route Planning (CRP) [32] algorithm is a variant of this approach. The hierarchical method takes advantage of the hierarchical structure of the road. Because of the existence of road hierarchy, the shortest path is not necessarily the optimal solution. This method introduces the weight of road into route planning. The Contraction Hierarchies (CH) [33] algorithm is a hierarchical method to create shortcuts to skip vertices with low importance. Bound-hop method is an ingenious method. By calculating the distance and time between two points in advance, the time-consuming of route planning is greatly reduced. When this method is directly applied to large-scale network, it is not feasible, because it will consume a lot of storage space. A bounded-hop algorithm is Transit Node Routing (TNR) [34], which uses distance tables on a subset of the vertices. At present, the route planning algorithm which combines the above methods is very mature and can be queried in milliseconds. The weighted vehicle routing problem (WVRP) has many applications, such as bus service, freight transportation by weight and hazardous waste collection. In [35], a mathematical model of WVRP is proposed, which corresponds to the two cases of receiving and delivering goods, and proposes an efficient heuristic method (RI-ILS) to solve WVRPs.

Path planning is a kind of short-term planning, which is used to find the path to avoid obstacles and meet the enhancement standards in the configuration space of given start and end points. Graph based path planners apply the same technology as route planners, such as A* algorithm, but the output is discrete path rather than continuous path, which may lead to unsmooth path. Yoon et al. [36] proposed a variation of A-star to compute the path. Sampling based method establishes the connectivity of configuration space by randomly sampling the paths in configuration space. Because this method is based on random sampling, the path obtained is not stable. Interpolating curve method fits the curve to a set of known points. The principle obstacle avoidance procedure is to interject the new crash free way, first deviate, ten return to the initial planning direction, and produce another way by fitting the bend to a group of new points. The path obtained by this method is smooth, but the amount of computation is high [37] use cubic spline curves for path planning. Recently, deep learning method is gradually applied to path planning. Convolutional neural network based on 3D can generate path from lidar point cloud and other inputs. Planning a path in impeded intersections was achieved in a simulation environment using deep reinforcement learning in [38]. The generalization of deep learning methods and the need to label data are the difficulties to be overcome.

Behavior selector is used to select the driving behavior, such as lane keeping, intersection processing, traffic light processing, etc. Driverless vehicles must deal with all kinds of road and urban traffic conditions in the practical scene. At present, the ability of behavior selector is limited, and the behavior selector which has not reached the function of level 4 has not appeared, but the related research has been carried out. Decision tree, finite state machine and heuristics are some traditional methods, which are used in limited and simple scenarios. Considering the uncertainty of intention and trajectory of moving objects, Markov decision process is applied to related tasks and achieves certain results. Wray et al. [39] proposed a comprehensive reasoning and behavior selection method to model car behavior and nearby cars as a set of discrete strategies. However, they believe that most of the drivers will follow the traffic rules in a normal and predictable way.

8 Datasets

Datasets are very important for autonomous driving because most algorithms and tools must be trained and tested before they are put into use. The common use is to test and verify these algorithms on the datasets with labels. With the gradual maturity of related algorithms, the training datasets supporting them are gradually increasing. Table 1 is a summary of autonomous driving datasets.

In addition to using data sets, camera simulation is also a method. Camera simulation can be used to test the perception module, such as building some scenes that are difficult to encounter or dangerous in the real world, or generating training data to reduce the need for manual annotation. GeoSim [40] combines graphics and neural networks to insert a new car into a given video or image. GeoSim can generate ultra-high-resolution video (4096×2160).

9 Transportation in Smart City

Transportation is one of the most important and challenging areas of smart city. Lin et al. [48] predicted that the future traffic mode will be the combination of smart traffic control and driverless cars. The goal of smart traffic control is to establish a system, so that all cars, traffic signs and control servers can share information and make correct decisions in a safe and optimized environment. The components of the smart transportation system can share all the necessary information among them, and then make the best decision through appropriate analysis. In this respect, there are many hotspots in the field of intelligent traffic control for autonomous cars [49]. Lee and Chiu [50] designed and implemented an intelligent traffic signal control (STSC) system based on V2X, which can reduce traffic congestion and improve

Table 1 Summary of datasets for training autonomous driving systems

Dataset	Problem space	Sensor	Size	Traffic condition
AMUSE [41]	SLAM	Omnidirectional camera, IMU, EgoData, GPS	1 TB, 7 clips	Urban
ApoiloScape [42]	2D/3D tracking 2D/3D object detection, segmentation, pedestrian detection, lane detection	Stereo and monocular cameras, GPS Lidar, IMU	200 K frame, 26 clips	Urban
Cityscapes [43]	Semantic understanding	Color stereo cameras	63 GB, 5 clips	Urban
KITTI [44]	3D tracking, 3D object detection, SLAM	Monocular cameras, IMU Lidar, GPS	180 GB	Urban, Rural
NuScenes [45]	3D tracking, 3D object detection	Radar, Lidar, EgoData, GPS, IMU, Camera	345 GB, 20 clips	Urban
Oxford [46]	3D tracking, 3D object detection, SLAM	Stereo and monocular cameras, GPS Lidar. IMU	23 TB, 133 clips	Urban, Highway
Udacity [47]	3D tracking, 3D object detection	Monocular cameras, IMU, Lidar, GPS, EgoData	220 GB	Rural

public transport efficiency. Zhu et al. [51] proposed a parallel transportation system of smart transportation system combined with artificial intelligent transportation system. In the smart transportation system, traffic flow prediction is helpful to analyze road conditions and feed-back traffic situation to managers and travelers in time. Chen et al. [52] proposes a traffic flow prediction method of city road network based on deep learning. Lv et al. [53] proposes a deep learning framework T-MGCN (Temporal Multi-Graph Convolutional Network) for traffic flow prediction in order to jointly model the spatial, temporal and semantic correlation of various features in road network.

In the concept of smart traffic control, path planning analysis is a challenging problem. Path planning in the environment of driverless vehicles is a complex big data problem, the traditional method based on single vehicle will not help [54]. The fundamental issue is to realize and keep up with the continuous connection between all components of the system. Considering the situation of driverless vehicles passing through streets and intersections, according to the traffic stream, all information gathered from vehicles and their destination and traffic limitations, suitable traffic control ought to be set up, and on this premise, fitting traffic light system ought to be coordinated. In future smart traffic control, a definitive objective is to eliminate traffic signs and limit time and distance as a result of driverless vehicles. At present, real-time analysis of a large number of different data collected from driverless vehicles is still a challenge. Strict delay request can be used to compensate [55] by introducing the analysis function near the driverless vehicle through the concept of edge computing. Solutions based on data fusion and modern artificial intelligence algorithms help to analyze data accurately and enhance decision-making ability. Specifically, the fusion of edge computing and artificial intelligence is an ideal solution to obtain real-time insight of vehicle data.

10 Deep Learning in Driverless Vehicles

In the past decade, deep learning has become the main technology behind many breakthroughs in computer vision, natural language processing and robotics. They also have a significant impact on today's driverless vehicles. Driverless vehicles are moving from laboratory development and testing to driving on public roads. Driverless vehicles have complex architecture and need to deal with many tasks, such as localization, perception, planning and so on. Although these tasks have made some progress under the traditional methods, the effect cannot reach the standard of practical use. In addition, different methods are usually used for different tasks, which increases the cost of system integration. The introduction of deep learning improves these problems to a great extent. With regard to driverless vehicles, deep learning technology can provide facilitation in terms of object detection, object tracking, Semantic segmentation and planning [18, 19, 21–28, 30, 38].

End to end deep learning combined with reinforcement learning for autonomous driving is very popular. The information to action mode of end-to-end system is closer to human driving habits. Modular system makes decision after understanding the whole scene, which involves many related problems. Comprehensive decision-making is a very difficult and complex thing. The end-to-end system leaves the whole thing to the neural network to do, so it doesn't need to make rules artificially, and the cost of the system is lower than that of the modular system. However, with the same information, there is more than one possible action. As in real life, different drivers make different decisions in the face of the same or similar scenes, so end-to-end deep learning is like an ill-posed problem. It can be predicted that deep learning will continue to play an important role in the field of driverless vehicles in the future.

11 Driverless Vehicle Cases

Due to the development of related technologies and the social demand for driverless vehicles, many traditional automobile manufacturers, new energy automobile manufacturers and even Internet companies begin to develop driverless vehicles. At

Fig. 4 Baidu Apollo GO driverless vehicle [56]

present, many automobile manufacturers have launched Level 2 driverless vehicles, such as Audi A6L, Mercedes Benz S-Class, Cadillac CT6, etc. Most level 2 vehicles can only achieve acceleration and deceleration in a single lane. At present, many automobile manufacturers have launched Level 2 driverless vehicles, such as Audi A6L, Mercedes Benz S-class, Cadillac CT6, etc. Most level 2 vehicles can only achieve acceleration and deceleration in a single lane. Tesla claims that its Tesla Model S has reached level 2.5, that is, on the basis of level 2, it has realized semi-automatic lane change—the driver judges the environment, determines that it can change lanes, and the vehicle changes lanes.

At present, there is no level 3 vehicle on the market. At present, there is no level 3 vehicle on the market, but many manufacturers have realized full autonomous driving, such as Baidu Apollo, Waymo, Pony.ai. Apollo [56] is an autonomous driving plan released by Baidu (Fig. 4), including open platform and enterprise solutions. Currently, Apollo has launched Apollo GO and Minibus driverless vehicles. In addition, there are smart transportation solutions such as Valet Parking, smart traffic signals and ACE. In addition, Baidu also launched an Apollo open platform. It is a high performance, flexible architecture which accelerates the development, testing, and deployment of autonomous vehicles [8]. Waymo [57] was originally a Google driverless vehicle project owned by Alphabet and is now independent. Waymo has launched passenger service Waymo One and freight service Waymo Via, among which Waymo One is in trial operation in the suburb of Phoenix. After a long-term open road test in urban area, Pony.ai [58] has accumulated a large number of complex and extreme scene data, and can safely and intelligently process these scenes to realize safe and reliable application of autonomous driving technology. Pony Alpha, Pony Alpha 2 and Pony Alpha X were introduced. PonyPilot is a self-driving travel service launched by Pony.ai in December 2018. It operates normally in Guangzhou and will be extended to Beijing, Shanghai, Fremont and Irvine in 2021.

12   Conclusion

Although the prospect of autonomous driving is attractive and a number of driverless vehicles have been launched, there are still many challenges. Several architecture models have been proposed, each of which has its own advantages and limitations. With the help of deep learning, localization and perception algorithm has a good improvement. If it is put into commercial use, it still lacks accuracy and efficiency. V2X network is still in its infancy, but due to the complex infrastructure, centralized cloud-based information management system has not been implemented. With the construction of smart city infrastructure, the application of driverless vehicles will have a good prospect.

It can be predicted that driverless vehicles will be able to develop with the promotion of deep learning, especially in the end-to-end architecture. There is still a lot of room for improvement in perception, planning and decision-making. With the construction of smart city technology facilities, especially communication facilities, V2X can bring far-reaching impact to driverless vehicles. Although the number and functions of driverless vehicles on the market are limited, with the progress of technology, higher level driverless vehicles can enter the market. Driverless vehicles will play an important role in urban transportation, improving the use of land resources and the quality of life.

References

1. Self-driving cars explained. https://www.ucsusa.org/resources/self-driving-cars-101. Accessed 6 Jun 2021
2. Yigitcanlar T (2016) Technology and the city: systems, applications and implications. Routledge, New York
3. Kim J, Moon Y-J, Suh I-S (2015) Smart mobility strategy in Korea on sustainability, safety and efficiency toward 2025. IEEE Intell Transp Syst Mag 7(4):58–67
4. Chen Q, Ozguner U, Redmill K (2004) Ohio State University at the 2004 DARPA grand challenge: developing a completely autonomous vehicle. IEEE Intell Syst 19(5):8–11
5. Buehler M, Iagnemma K, Singh S (2009) The DARPA urban challenge: autonomous vehicles in city traffic 1st, pp 626–626
6. SAE. Taxonomy and definitions for terms related to driving automation systems for on-road motor vehicles. SAE J3016. 2021. Tech Rep
7. Gold C, Körber M, Lechner D, Bengler K (2016) Taking over control from highly automated vehicles in complex traffic situations the role of traffic density. Hum Factors 58(4):642–652
8. Baidu. Apollo auto. https://github.com/ApolloAuto/apollo. Accessed 8 Jun 2021
9. Abboud K, Omar HA, Zhuang W (2016) Interworking of DSRC and cellular network technologies for V2X communications: a survey. IEEE Trans Veh Technol 65(12):9457–9470
10. Hecker S, Dai D, Gool LV (2018) End-to-End learning of driving models with surround-view cameras and route planners. In: Proceedings of the european conference on computer vision (ECCV),vol 11211, pp 449–468
11. Chen J, Li SE, Tomizuka M (2021) Interpretable end-to-end urban autonomous driving with latent deep reinforcement learning. IEEE Trans Intell Transp Syst 1–11

12. Akai N, Morales LY, Takeuchi E, Yoshihara Y, Ninomiya Y (2017) Robust localization using 3D NDT scan matching with experimentally determined uncertainty and road marker matching. In: 2017 IEEE intelligent vehicles symposium IV, pp 1356–1363
13. Chen C, Wang B, Lu CX, Trigoni N, Markham A (2020) A survey on deep learning for localization and mapping: towards the age of spatial machine intelligence. ArXiv Preprint ArXiv:2006.12567
14. Redmon J, Divvala S, Girshick R, Farhadi, A (2016) You only look once: unified, real-time object detection. In: 2016 IEEE conference on computer vision and pattern recognition (CVPR) (pp. 779–788).
15. Liu W, Anguelov D, Erhan D, Szegedy C, Reed SE, Fu CY, Berg AC (2016) SSD: single shot multibox detector. In: 14th European conference on computer vision, ECCV 2016, pp 21–37
16. YOLOv5. https://github.com/ultralytics/yolov5. Accessed 8 Jun 2021
17. Ren S, He K, Girshick R, Sun J (2017) Faster R-CNN: towards real-time object detection with region proposal networks. IEEE Trans Pattern Anal Mach Intell 39(6):1137–1149
18. Lee Y, Park J (2020) CenterMask: real-time anchor-free instance segmentation. In: 2020 IEEE/CVF conference on computer vision and pattern recognition (CVPR), pp 13906–13915
19. Mohan R, Valada A (2021) EfficientPS: efficient panoptic segmentation. Int J Comput Vision 129(5):1551–1579
20. Dong G, Yan Y, Shen C, Wang H (2020) Real-time high-performance semantic image segmentation of urban street scenes. IEEE Trans Intell Transp Syst, 1–17
21. Charles RQ, Su H, Kaichun M, Guibas LJ (2017) PointNet: deep learning on point sets for 3d classification and segmentation. In: 2017 IEEE conference on computer vision and pattern recognition (CVPR), pp 77–85
22. Kuang H, Wang B, An J, Zhang M, Zhang Z (2020) Voxel-FPN: multi-scale voxel feature aggregation for 3D object detection from LIDAR point clouds. Sensors 20(3):704
23. Xie Q, Lai YK, Wu J, Wang Z, Zhang Y, Xu K, Wang J (2020) MLCVNet: multi-level context VoteNet for 3D object detection. In: 2020 IEEE/CVF conference on computer vision and pattern recognition (CVPR), pp 10447–10456
24. Brazil G, Liu X (2019) M3D-RPN: monocular 3D region proposal network for object detection. In: 2019 IEEE/CVF international conference on computer vision (ICCV), pp 9287–9296
25. Scheidegger S, Benjaminsson J, Rosenberg E, Krishnan A, Granstrom, K (2018) Mono-Camera 3D multi-object tracking using deep learning detections and PMBM filtering. In: 2018 IEEE intelligent vehicles symposium (IV), pp 433–440
26. Naiden A, Paunescu V, Kim G, Jeon B, Leordeanu M (2019) Shift R-CNN: deep monocular 3D object detection with closed-form geometric constraints. In: 2019 IEEE international conference on image processing (ICIP), pp 61–65
27. Li B, Wu W, Wang Q, Zhang F, Xing J, Yan J (2019) SiamRPN++: evolution of siamese visual tracking with very deep networks. In: 2019 IEEE/CVF conference on computer vision and pattern recognition (CVPR), pp 4282–4291
28. Danelljan M, Gool LV, Timofte R (2020) Probabilistic regression for visual tracking. In: 2020 IEEE/CVF conference on computer vision and pattern recognition (CVPR), pp 7183–7192
29. Song X, Chen, K, Li X, Sun J, Hou B, Cui Y, Wang Z et al (2020) Pedestrian trajectory prediction based on deep convolutional LSTM network. IEEE Trans Intell Transp Syst, 1–18
30. Caltagirone L, Bellone M, Svensson L, Wahde M (2019) LIDAR–camera fusion for road detection using fully convolutional neural networks. Robot Auton Syst 111:125–131
31. Hilger M, Köhler E, Möhring RH, Schilling H (2006). Fast Point-to-point shortest path computations with arc-flags. The shortest path Problem, pp 41–72
32. Delling D, Goldberg AV, Pajor T, Werneck RF (2017) Customizable route planning in road networks. Transp Sci 51(2):566–591
33. Geisberger R, Sanders P, Schultes D, Vetter C (2012) Exact routing in large road networks using contraction hierarchies. Transp Sci 46(3):388–404
34. Arz J, Luxen D, Sanders P (2013) Transit node routing reconsidered. In: Bonifaci V (ed) Experimental algorithms : 12th international symposium ; proceedings, SEA 2013, Rome, Italy, 5–7June 2013. pp 55–66

35. Wang X, Shao S, Tang J (2020) Iterative local-search heuristic for weighted vehicle routing problem. IEEE Trans on Intell Transp Syst, 1–11
36. Yoon S, Yoon S-E, Lee U, Shim DH (2015) Recursive path planning using reduced states for car-like vehicles on grid maps. IEEE Trans Intell Transp Syst 16(5):2797–2813
37. Hu X, Chen L, Tang B, Cao D, He H (2018) Dynamic path planning for autonomous driving on various roads with avoidance of static and moving obstacles. Mech Syst Signal Process 100(100):482–500
38. Isele D, Rahimi R, Cosgun A, Subramanian K, Fujimura K (2018) Navigating occluded intersections with autonomous vehicles using deep reinforcement learning. In: 2018 IEEE international conference on robotics and automation (ICRA), pp 2034–2039
39. Wray KH, Witwicki SJ, Zilberstein S (2017) Online decision-making for scalable autonomous systems. In: IJCAI'17 proceedings of the 26th international joint conference on artificial intelligence, pp 4768–4774
40. Chen Y, Rong F, Duggal S, Wang S, Yan X, Manivasagam S, Urtasun R et al (2021) GeoSim: realistic video simulation via geometry-aware composition for self-driving. ArXiv Preprint ArXiv:2101.06543
41. Koschorrek P, Piccini T, Oberg P, Felsberg M, Nielsen L, Mester R (2013) A multi-sensor traffic scene dataset with omnidirectional video. In: 2013 IEEE conference on computer vision and pattern recognition workshops, pp 727–734
42. Huang X, Cheng X, Geng Q, Cao B, Zhou D, Wang P, Yang R et al (2018) The apolloscape dataset for autonomous driving. In: 2018 IEEE/CVF conference on computer vision and pattern recognition workshops (CVPRW), pp 954–960
43. Cityscapes. Cityscapes data collection. http://www.cityscapes-dataset.com/. Accessed 8 Jun 2021
44. Geiger A, Lenz P, Stiller C, Urtasun R (2013) Vision meets robotics: The KITTI dataset. The Int J Robot Res 32(11):1231–1237
45. Caesar H, Bankiti V, Lang AH, Vora S, Liong VE, Xu Q, Beijbom O et al (2020) nuScenes: A multimodal dataset for autonomous driving. In: 2020 IEEE/CVF conference on computer vision and pattern recognition (CVPR), pp 11621–11631
46. Maddern W, Pascoe G, Linegar C, Newman P (2017) 1 year, 1000 km: The Oxford RobotCar dataset. Int J Robot Res, 36(1): 3–15
47. Udacity. Udacity Data Collection. http: //academictorrents.com/collection/self-driving-cars. Accessed 8 Jun 2021
48. Lin Y, Wang P, Ma M (2017) Intelligent transportation system(ITS): concept, challenge and opportunity. In: 2017 IEEE 3rd international conference on big data security on cloud (BigDataSecurity), ieee international conference on high performance and smart computing, (HPSC) and IEEE international conference on intelligent data and security (IDS), pp 167–172
49. Nigon J, Glize E, Dupas D, Crasnier F, Boes J (2016) Use cases of pervasive artificial intelligence for smart cities challenges. In: 2016 Intl IEEE conferences on ubiquitous intelligence & computing, advanced and trusted computing, scalable computing and communications, cloud and big data computing, internet of people, and smart world congress (UIC/ATC/ScalCom/CBDCom/IoP/SmartWorld), pp 1021–1027
50. Lee WH, Chiu CY (2020) Design and implementation of a smart traffic signal control system for smart city applications. Sensors 20(2):508
51. Zhu F, Lv Y, Chen Y, Wang X, Xiong G, Wang F-Y (2020) Parallel transportation systems: toward iot-enabled smart urban traffic control and management. IEEE Trans Intell Transp Syst 21(10):4063–4071
52. Chen C, Liu Z, Wan S, Luan J, Pei Q (2020). Traffic flow prediction based on deep learning in internet of vehicles. IEEE Trans Intell Transp Syst, 1–14
53. Lv M, Hong Z, Chen L, Chen T, Zhu T, Ji S (2020) Temporal multi-graph convolutional network for traffic flow prediction. IEEE Trans Intell Transp Syst, 1–12
54. Toole JL, Colak S, Sturt B, Alexander LP, Evsukoff AG, González MC (2015) The path most travelled: travel demand estimation using big data resources. Transp Res Part C-Emerging Technol 58:162–177

55. Yuan Q, Zhou H, Li J, Liu Z, Yang F, Shen XS (2018) Toward efficient content delivery for automated driving services: an edge computing solution. IEEE Network 32(1):80–86
56. Apollo. Baidu. https://apollo.auto/. Accessed 10 Jun 2021
57. Waywo. https://waymo.com/. Accessed 10 Jun 2021
58. Pony.ai. https://www.pony.ai/. Accessed 10 Jun 2021
59. Grigorescu SM, Trasnea B, Cocias TT, Macesanu G (2020) A survey of deep learning techniques for autonomous driving. J Field Robot 37(3):362–386
60. Azgomi HF, Jamshidi M (2018) A brief survey on smart community and smart transportation. In: 2018 IEEE 30th international conference on tools with artificial intelligence (ICTAI), pp 932–939

A Mechanism to Maintain Node Integrity in Decentralised Systems

Monjur Ahmed and Md. Mahabub Alam Bhuiyan

Abstract This paper presents a mechanism to maintain node integrity of decentralised systems consisted of nodes or microservices. The mechanism addresses total counts of nodes or micro-services in a decentralised systems. The proposed mechanism is a further development of a decentralised security model named Ki-Ngā-Kōpuku. Decentralised systems where the total number of active nodes matters to maintain the integrity of the system, it is important for the nodes to form a closed group among the valid nodes; as well as be aware of any node that becomes inactive or of any foreign node that may want to join the closed group of nodes. The nodes of a decentralised system need to be aware of the members of other nodes in the same group of nodes, and need to be able to raise alarm in the event of any changes in total number of nodes for whatever is the reason. The mechanism presented in this paper is a proposed solution to maintain integrity of a decentralised system in terms of number of nodes. The mechanism relies on beacon transmission to achieve its goal. The concept of the proposed mechanism applies to other contexts, for example, where a system consists of multiple micro-services.

1 Introduction

Smart cities and Industry 4.0 are two emerging trends. Industry 4.0 enables and incorporates IoT and other internet enabled systems to be a common picture in industrial automation, contexts and operations. Industry 4.0 is highly automatised [1, 2]. The emerging trend of industry 4.0 incorporates cybersecurity concerns. The example of Stuxnet [3–5] informs us that the industrial systems are prone to cyber-attacks, and cyber-resilience in industrial automation context cannot be undermined. Similarly, the above equally applies to smart cities since Industry 4.0 and the concept of smart

M. Ahmed (✉) · Md. M. A. Bhuiyan
Waikato Institute of Technology, Hamilton, New Zealand
e-mail: monjur.ahmed@wintec.ac.nz

Md. M. A. Bhuiyan
e-mail: mdxbhu21@student.wintec.ac.nz

© Springer Nature Switzerland AG 2022
R. Jiang et al. (eds.), *Big Data Privacy and Security in Smart Cities*,
Advanced Sciences and Technologies for Security Applications,
https://doi.org/10.1007/978-3-031-04424-3_4

cities are inter-related to some good extent. Highly integrated IoT and embedded systems within the context of a smart city could be a prime target for adversaries for numerous reasons. Cybersecurity threats in Industry 4.0 and smart city contexts are probably the most crucial concerns. One of the approaches to combat cyber-attack is to make a system decentralised with no single or centralised core. For example, a decentralised and distributed security model is proposed in [6] that consists of several nodes with no single core of the system. Such node-based decentralised systems, and any other system that has multiple components as part of it require a mechanism in place to keep track of the nodes or components that it is made up of.

Decentralisation is achieved through collaborative nodes [7–9]. A decentralised system may tend to eliminate a single core of a system by componentising itself. Such a decentralised system is made up of a number of different entities (e.g., nodes) that collectively make the whole of the system but a single part (or node) of the system does not constitute any significant portion of the system. A single node of the above decentralised system does not represent the system from a holistic point-of-view, yet it is an essential part for the system to function. This tends to make a decentralised system having no single core of the system. There is no central controlling node to administer the system, but all nodes collaboratively and collectively serve as a system. In such a decentralised system, all nodes are equally powerful when work collaboratively, yet no node is powerful when on its own. The above is the core principle of Ki-Ngā-Kōpuku—a decentralised and distributed security model for Cloud Computing [6]. However, a member node of such decentralised system, if acts with malicious intent and trusted by other nodes, may devastate the integrity of the system. To avoid such loss of integrity (and eventually to avoid any system level compromise), it is important for the nodes of a decentralised system to be aware of any changes in total number of nodes, for example if a becomes inactive or a new nodes tries to be part of the node-family of a decentralised system. Failure to do so may lead to a non-integrated decentralised system. A non-integrated decentralised system may have foreign nodes becoming member of the system and subsequently poising the system; or a native node that becomes inactive, zombied and then made active to join the family of the nodes of a decentralised system to eventually carry out malicious actions.

This paper presents the concept of a mechanism that aims to prevent the aforementioned anomaly to maintain the integrity of the system. The mechanism presented is a further development for a decentralised, distributed security model named Ki-Ngā-Kōpuku [6]. Node integrity is a crucial aspect for systems like Ki-Ngā-Kōpuku or any other decentralised node-based systems, which is the motivation to propose the mechanism presented in this paper. The rest of the paper is structured as follows: Related Study explores relevant research on the same track. The mechanism is described and illustrated in the Proposed Mechanism section followed by Discussions and Future Developments.

2 Related Study

Since the proposed mechanism is an enhancement of a previously proposed decentralised security model [6] to combat cyberthreat, in this section we focus on existing research on decentralised systems within industrial control and automation context. Decentralised system is argued to bring better integrity in industrial automation and in other areas. For example, Hao et al. [10] proposes a decentralised approach in communications engineering, while Melekhova et al. [11] proposes decentralised system for large-scale autonomic systems. Fanaeepour and Kangavari [7] uses decentralised approach to detect malicious nodes in sensor networks. [12] proposes an approach to handle node churn in decentralised system, since such churn may lead to performance degradation. Hurst et al. [13] state that a decentralised approach may bring better resilience in IoT systems. Decentralised approach for better cybersecurity is also argued in [14].

A system consisting of decentralised node may effectively combat cyber threats. For example, Raje et al. [15] propose a decentralise firewall to detect malware. The use of decentralised approach for better security is also described in [16]. Wood [17] describes a decentralised system and claims it to be a secure generalised transaction ledger system. Khnan and Lawal [18] propose a decentralised framework to secure Internet-of-Things (IoT). Research incorporating decentralised systems and its nodes address various collaboration and coordination aspects of the nodes in a decentralised system, such examples are found in [19–21]. Integrity of a decentralised system is complex yet a pre-requisite. The integrity related aspects of decentralised or other systems are considered by researchers, for example in [22–25].

Some decentralised systems consist exclusively of nodes that are operators and there is no additional infrastructure [26]. The nodes present in the decentralised system verification with public auditability cannot avoid in the Third-Party Auditors (TPAs) [8]. Studies have stated that in a decentralised system node integrity is essential as this allows scalability and increase in system capability [9]. The decentralised system with integrity offers higher security to the system processes. Additionally, node integrity present in the decentralised system offers higher resilience [27]. Node integrity in decentralised system are essential for system security, service availability and effective decision-making capability [28]. Thus, node integrity is essential to reduce system control risk and increases system efficiency.

3 Proposed Mechanism

We first revisit (before outlining the proposed mechanism) Ki-Ngā-Kōpuku architecture since the proposed mechanism complements Ki-Ngā-Kōpuku as well as any other system built upon Ki-Ngā-Kōpuku or similar principle. Figure 1 illustrates the application componentisation principle of Ki-Ngā-Kōpuku.

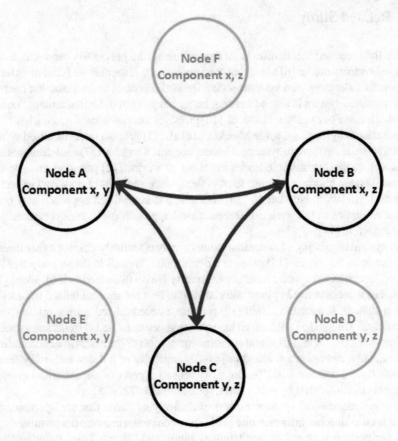

Fig. 1 Ki-Ngā-Kōpuku principle (Adapted from [6])

Ki-Ngā-Kōpuku divides an application into multiple components (componenti-sation) and places the components in different nodes in such a manner that no single node contains all components of an application [6]. A node also has its redundant copies that gets activated in case any active node disappears. Referring to Fig. 1, The system deploys an application with three components x, y, z that reside in six nodes A, B, C, D, E and F. Nodes A, B, C, are active nodes and nodes D, E, F are redundant nodes. The system has no core and all nodes collectively defines the whole system. For any reason, if node A becomes inactive, the system needs to detect it and at the same time, node A, B and C should ideally have no knowledge on the total number of nodes (three in this case) that makes the whole system. The scope of the mechanism presented in this paper is the detection of a node being inactive.

To maintain node integrity in terms of total number of nodes, each node (base node) will tie-up with n number of other nodes (tie-up nodes). In the illustrative example in Figure 2, we assume n = 2. In other words, each node will tie-up with two other nodes. As illustrated in Fig. 2, node C is the base node and the other two

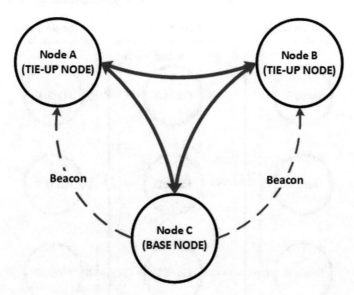

Fig. 2 Base node and tie-up node

nodes, node A and node B are tie-up nodes. In the same was, when A is the base node, node B and node C are tie-up nodes for node A. All nodes act as both the base node and tie-up node.

The base node will intermittently communicate with tie-up nodes and the communication from a base node to its tie-up nodes must not be synchronous. For example, if the communication threshold is two seconds (T1 and T2), the base node will send 'beacon' to one tie-up node at T1, and the other beacon to the other tie-up node on T2. The communication among base node and tie-up nodes, within the context of the mechanism, is only to send beacons (by base node) to inform tie-up nodes that it (the base node) is alive. A base node may essentially act as a tie-up node for another base node. In other words, each nodes of a decentralised system will act as both base-node and tie-up node. As the base-node, a node transmits its existence to its tie-up nodes, and receives transmission from other node(s) for which it is acting as a tie-up node. This is illustrated in Fig. 3, a sends beacon to its tie-up nodes (b and c) but not at the same time. The sequence of beacon transmission by base node to tie-up nodes is insignificant. Figure 3 illustrates the above, where step 1 and step 2 are indefinitely repetitive, asynchronous and not prioritised in any order.

Assuming,

P = the proposed mechanism

x = a base node

y = tie-up nodes

N = natural number

n = number of tie-up nodes.

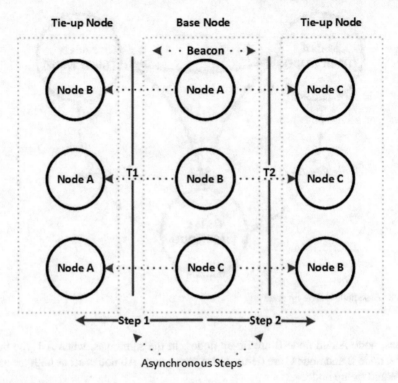

Fig. 3 Beacon transmission among base nodes and tie-up nodes

The mechanism and the relationship between a base node and its tie-up nodes are noted in (1).

$$P \rightarrow \forall x \exists y \{(x \notin y) \wedge (n \in N : n > 1)\} \tag{1}$$

The non-transmission of a timely beacon destined to tie-up nodes is a possible indication that the base node is unavailable for whatever the reason is. The beacon itself is not a mere signal of a node being alive since any node may impersonate such beacon. The metaphorical beacon in this context could be a message passing mechanism to verify the authenticity of a base node by its tie-up nodes. Any valid approach to verify the authenticity can be adapted to achieve this. One such example is found in [6] where nodes use a combination of their own checksum and asymmetric encryption to verify the authenticity of node there are coupled into (e.g., the coupling among base node and tie-up nodes). The process of authentication and verification is outside of the scope of this paper and is planned to explore in future research.

4 Discussions and Future Development

The mechanism aims to achieve the detection of lost node in a decentralised system without requiring to learn the total number of nodes. When a node fails to send beacons to its tie-up nodes, the tie-up nodes take it as a possibility of the base node being unavailable or compromised. The absence of a timely beacon destined to tie-up nodes is a possible indication of the node becoming unavailable. Such indication leads to a system wide knowledge that a node has become unavailable. While a node does not have any knowledge on the total count of nodes of a system (to avoid the scope of learning the context of a system by an adversary), the tie-up nodes are able to detect the 'death' of a node and can later prevent the same node from being part of the system or a foreign node impersonating the dead node. In systems made upon Ki-Ngā-Kōpuku principle, each node has its redundant copies, and there is no need to recover a dead node. According to Ki-Ngā-Kōpuku principle, a suspicious node can simply be discarded, and a redundant copy of the node may subsequently become active; this eliminates the requirement of recovering a lost node and accepting any residual risks involved as a result of such recovery. The novelty of the mechanism stands within its ability of maintain node integrity of a decentralised system without having the knowledge of total number of nodes.

The proposed mechanism is a further development of the previous research and focuses only on detecting whether all nodes in a decentralised node are active. As mentioned earlier, future developments of the proposed mechanism are to develop the beaconing process of the nodes, as well as explore the existing options that may be adapted for beaconing. Further investigation is planned to find out how a decentralised system may combat cyber threats in industry 4.0 context and how the proposed mechanism may help decentralised systems to strengthen the maintenance of overall integrity.

5 Conclusions

The concept of the proposed mechanism can be applied to any context where a member of a set needs to identify whether all members are present, without knowing the total number of members in the set. Using the mechanism proposed, the members of a set (of some entity) can achieve the above only by working collectively, and without the need to know the total number of members of the set they belong to. Decentralised systems tend to incorporate massive processing. The proposed mechanism will add further processing demand on an existing infrastructure of a decentralised system. However, the trade-off is between the processing overheads incorporated with the proposed mechanism and the level of integrity it may add to a decentralised system.

References

1. Lasi H, Fettke P, Kemper H, Feld T, Hoffmann M (2014) Industry 4.0. Bus Inf Syst Eng 6(4):239–242. https://doi.org/10.1007/s12599-014-0334-4
2. Haseeb M, Hussain H, Ślusarczyk B, Jermsittiparsert K (2019) Industry 4.0: a solution towards technology challenges of sustainable business performance. Soc Scie 8(5):154. https://doi.org/10.3390/socsci8050154
3. Farwell J, Rohozinski R (2011) Stuxnet and the future of cyber war. Survival 53(1):23–40. https://doi.org/10.1080/00396338.2011.555586
4. Trautman LJ, Ormerod PC (2017) Industrial cyber vulnerabilities: lessons from Stuxnet and the Internet of Things. U Miami L Rev 72:761
5. Falliere N, Murchu LO, Chien E (2011) W32. stuxnet dossier. In: White paper, symantec corp., security response, vol 5, no 6, p 29
6. Ahmed M (2018) Ki-Ngā-Kōpuku: a decentralised, distributed security model for cloud computing. Auckland University of Technology
7. Fanaeepour M, Kangavari M (2010) Malicious node detection system in wireless sensor networks: a decentralised approach. Int J Internet Technol Secured Trans 2(12):88. https://doi.org/10.1504/ijitst.2010.031473
8. Dong P et al (2020) Edge computing based healthcare systems: enabling decentralized health monitoring in internet of medical things. IEEE Netw 34(5):254–261. https://doi.org/10.1109/MNET.011.1900636
9. Bengio Y et al (2021) Inherent privacy limitations of decentralized contact tracing apps. J Am Med Inf Assoc 28(1):193–195. https://doi.org/10.1093/jamia/ocaa153
10. Hao Y, Duan Z, Chen G (2018) Decentralised fixed modes of networked MIMO systems. Int J Control 91(4):859–873. https://doi.org/10.1080/00207179.2017.1295318
11. Melekhova O, Malenfant J, Mescheriakov R, Chueshev A (2018) A decentralised solution for coordinating decisions in large-scale autonomic systems. In: MATEC web of conferences, vol 161, p 03024. https://doi.org/10.1051/matecconf/201816103024
12. Chen Y, Zhao G, Li A, Deng B, Li X (2009) Handling node churn in decentralised network coordinate system. IET Commun 3(10):1578. https://doi.org/10.1049/iet-com.2008.0671
13. Hurst W, Shone N, El Rhalibi A, Happe A, Kotze B, Duncan B (2017) Advancing the micro-CI testbed for IoT cyber-security research and education. Cloud Comput 2017:139
14. Waller A, Craddock R (2011) Managing runtime re-engineering of a System-of-Systems for cyber security. In: 2011 6th International conference on system of systems engineering, pp 13–18. https://doi.org/10.1109/SYSOSE.2011.5966566
15. Raje S, Vaderia S, Wilson N, Panigrahi R (2017) Decentralised firewall for malware detection. In: 2017 International conference on advances in computing, communication and control (ICAC3), pp 1–5. https://doi.org/10.1109/ICAC3.2017.8318755
16. Li H, Venugopal S (2013) Efficient node bootstrapping for decentralised shared-nothing key-value stores. Middleware 2013:348–367. https://doi.org/10.1007/978-3-642-45065-5_18
17. Wood G (2014) Ethereum: a secure decentralised generalised transaction ledger. In: Ethereum Project yellow paper, vol 151, no 2014, pp 1–32
18. Khan M, Lawal I (2020) Sec-IoT: a framework for secured decentralised IoT using blockchain-based technology. In: Proceedings of fifth international congress on information and communication technology, pp 269–277. https://doi.org/10.1007/978-981-15-5856-6_27
19. Tejaswini B, Ashwini S, Shilpashree S (2018) Protected Data Reclamation for decentralised disruption-forbearing In wireless sensor network. Int J Eng Res Technol (IJERT)
20. Manfredi S, Di Tucci E A decentralised topology control to regulate global properties of complex networks. Eur Phys J B 90(4):https://doi.org/10.1140/epjb/e2017-70586-9
21. Xu Y, Huang Y (2020) An n/2 byzantine node tolerate blockchain sharding approach. In: Proceedings of the 35th annual ACM symposium on applied computing. https://doi.org/10.1145/3341105.3374069
22. Lin C, Shen Z, Chen Q, Sheldon F (2017) A data integrity verification scheme in mobile cloud computing. J Netw Comput Appl 77:146–151. https://doi.org/10.1016/j.jnca.2016.08.017

23. Spathoulas G et al (2018) Towards reliable integrity in blacklisting: facing malicious IPs in GHOST smart contracts. Innov Intell Syst Appl (INISTA) 2018:1–8. https://doi.org/10.1109/INISTA.2018.8466327
24. Schiffman J, Moyer T, Shal C, Jaeger T, McDaniel P (2009) Justifying integrity using a virtual machine verifier. In: Annual computer security applications conference, pp 83–92. https://doi.org/10.1109/ACSAC.2009.18
25. Altulyan M, Yao L, Kanhere S, Wang X, Huang C (2020) A unified framework for data integrity protection in people-centric smart cities. Multimedia Tools Appl 79(7–8):4989–5002. https://doi.org/10.1007/s11042-019-7182-7
26. Ezéchiel K, Ojha S, Agarwal R (2020) A new eager replication approach using a non-blocking protocol over a decentralized P2P architecture. Int J Distrib Syst Technol 11(2):69–100. https://doi.org/10.4018/ijdst.2020040106
27. Ghanati G, Azadi S (2020) Decentralized robust control of a vehicle's interior sound field. J Vibr Control 26(19–20):1815–1823. https://doi.org/10.1177/1077546320907760
28. Sumathi M, Sangeetha S (2020) Blockchain based sensitive attribute storage and access monitoring in banking system. Int J Cloud Appl Comput 10(2):77–92. https://doi.org/10.4018/ijcac.2020040105

Incident Detection System for Industrial Networks

Karel Kuchar, Eva Holasova, Radek Fujdiak, Petr Blazek, and Jiri Misurec

Abstract Modbus/TCP is one of the most used industrial protocol, but this protocol is unsecured and does not implement encryption of communication or authentication of the clients. Therefore, this paper is focused on the techniques of incident detection in Modbus/TCP communication, but it is possible to implement the proposed solution on different protocols. For this purpose, a Modbus Security Module was created. This module can sniff specific network traffic, parse particular information from the communication packets, and store this data into the database. The databases use PostgreSQL and are placed on each master and slave stations. The data stored in each database is used for incident detection. This method represents a new way of detecting incidents and cyber-attacks in the network. Using a neural network (with an accuracy of 99.52%), machine learning (with an accuracy of 100%), and database comparison, it is possible to detect all attacks targeting the slave station and detect simulated attacks originating from master or non-master station. For additional database security of each station, an SSH connection between the databases is used. For the evaluation of the proposed method, the IEEE dataset was used. This paper also presents a comparison of machine learning classifiers, where each classifier has adjusted parameters. A mutual comparison of machine learning classifiers (with or without memory parameter) was done.

1 Introduction

IT (Information Technology) and OT (Operational Technology) are increasingly interconnected. The main purpose of IT from the perspective of the CIA triad is Confidentiality, Integrity, and Availability. On the other hand, the main purposes of OT and industrial networks are Availability and Integrity, and Confidentiality was not necessarily needed (in time of building an industrial network). Moreover, OT and

K. Kuchar · E. Holasova · R. Fujdiak (✉) · P. Blazek · J. Misurec
Department of Telecommunications, Faculty of Electrical Engineering and Communication,
Brno University of Technology, Technicka 12, Brno 616 00, Czech Republic
e-mail: fujdiak@vut.cz

© Springer Nature Switzerland AG 2022   83
R. Jiang et al. (eds.), *Big Data Privacy and Security in Smart Cities*,
Advanced Sciences and Technologies for Security Applications,
https://doi.org/10.1007/978-3-031-04424-3_5

industrial networks often add another parameter as supervision or control. Industrial networks were designed for durability, reliability, and maintainability. Due to the fast digitalization, process automation, and remote control of industrial networks, when we can talk about smart grids, Industry 4.0, IoT, and so on, there is great pressure on security in these technologies. This brings many new challenges in the standards and technologies. For this purpose, new standards were created such as NIST 800-82 [1], IEC-62443 (ISA-99) [2], NERC CIP [3] and others. Industrial networks build these days already emphasize security and are taken into account when choosing the industrial protocol and when designing the structure of the entire network (according to which the individual devices are also selected). We provide the solution for cases, when the "old" industrial network uses an unsecured version of the protocol, for security incident detection without changes needed in the current network. The proposed solution uses Security Modules to provide additional security. The solution uses a database and neural network or machine learning classifiers for the detection of many cyber-attacks in the network in two Phases approach. The mentioned Security Module also enables data recovery and legitimate stations detection that are under the control of an attacker. This paper follows on from paper [4], where the design of the security module was proposed and provides a comparison of possible methods for incident detection in Phase 2 of the Security Module using machine learning classifiers.

2   Related Works

Many recent research papers are focusing on machine learning algorithms and neural networks for incident detection in industrial networks. Beaver et al. [5] are focusing on the detection of malicious communication in SCADA networks. They use and mutually compare six machine learning classifiers and also use a publicly available dataset. However, the reached accuracy using the selected model is not mentioned, also machine learning classifiers are not closer specified. In paper [6], Mubarak et al. focused on cyber-attack detection using machine learning. The paper compares eight classifiers, however, the authors do not use publicly available dataset, the specific settings of individual machine learning classifier is also not mentioned as the reached accuracy. Joshi et al. [7] focused on a semi-supervised approach for the detection of SCADA attacks. Their paper reached an accuracy of 100% on the publicly available dataset. In comparison to the [7] our papers provide adjusted parameters of machine learning classifiers comparison, the overall approach to solving security incidents with the possibility of data recovery, also uses data obtained from individual stations, and cover wider are of cyber-attacks. Alhaidari et al. [8] mutually compare three machine learning models with an accuracy of 99.998% on the publicly available dataset. However, our paper is focusing on a wider area of machine learning classifiers with many adjusted parameters and provides the chain of incident detection techniques. Bulle et al. [9] uses three machine learning models but the comparison of models is omitted, the models reached 92.77%. However, our

Table 1 Related works comparison

Citation	Year	ML classifiers	Adjusted parameters	Public dataset	Accuracy	Overall solution	Machine state parameter
[5]	2014	6	No	Yes	Not mentioned	No	No
[6]	2021	8	No	No	Not mentioned	No	No
[7]	2020	1	No	Yes	100%	No	No
[8]	2019	3	No	Yes	99.9998%	No	No
[9]	2020	3	No	No	92.77%	No	No
[10]	2019	2	No	No	99.58%	No	No
This paper	–	7	Yes	Yes	63.86–100%	Yes	Yes

paper uses a publicly available dataset and provides an overall solution for an incident detection system. Perez et al. [10] reached the highest accuracy of 99.58% and compared two machine learning classifiers. In comparison to the paper [10], our paper uses a publicly available dataset, provide a machine learning classifier with adjusted parameters comparison. Table 1 mutually compares selected related works in the number of used machine learnings classifiers, adjusted machine learning classifier's parameters usage, public available dataset as input data, reached accuracy, overall solution (providing chain of steps for incident detection), and machine state parameter (if used).

As might be visible from Table 1, our work focuses on the comparison of seven machine learning classifiers, where each model has modified parameters. These models are trained and validated on a publicly available dataset. Trained models are created to detect security incidents in multiple categories (one model for all categories). The work also provides an overall solution to security incidents in industrial networks and our solution does not require interventions in the existing network.

3 Industrial Protocol Modbus/TCP

To demonstrate the function of our developed solution for incident detection techniques, the Modbus protocol [11, 12] was chosen. This industrial protocol is one of the most widely used industrial protocols [13, 14]. The protocol distinguishes between two types of communication devices, master and slave devices. Communication is of the client/server, challenge/response type, where the communication is always controlled by the master (client) device. The master station queries the values stored in the slave device, or changes/controls them. The master/slave station operates on level 1 of the Purdue model [15], the connected actuators/sensors belong to

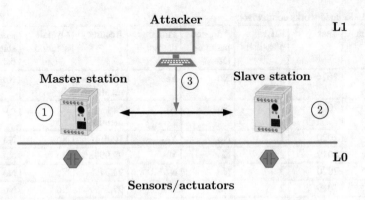

Fig. 1 Threat model

level 0. The industrial protocol Modbus uses function codes to specify the operation required by the master station. Each function code defines specific parameters that are transmitted in the message body. This industrial protocol, like most industrial protocols, was designed for closed network communication, completely separate from "regular" IT networks. For this reason, it lacks the implementation of security measures, such as authentication or encryption. In the event that IT and OT networks are interconnected, these networks are exposed to security risks and additional measures must be implemented to ensure the security or to inspect communications [16–18].

Even though the secure version of Modbus/TCP has been released, the unsecured version is still frequently used [19]. In order to upgrade the industrial protocol to a secure version, it is necessary to shut down the network, purchase new equipment, and do other steps that are not always available/possible. Therefore, this article focuses on the insecure version of the Modbus/TCP protocol and suggests a possible solution (chain of procedures) by which security incidents can be detected and mitigated.

To design a detection approach, it is necessary to detect attack vectors that a potential attacker can use to compromise the system. Figure 1 shows the threat model which is used in this paper, it might be also marked as attacks vectors. This paper contains attacks on master and slave devices (1, 2), where detection was designed using the Security Module (Phase 2). And attacks targeting transmission medium, where detection was designed using the Security Module (Phase 1). From the IT point of view, vectors are detected: attack on the master station (1), attack on the slave station (2), and an attack on the transmission medium (3). In this case, the physical attack on the devices themselves, as well as on the connected sensors and actuators, is not taken into account. Only attacks from the L1 layer and higher are considered. An overview of attack vectors, with a possible impact and the proposed detection, is shown in Table 2.

Table 2 Major Modbus/TCP attack vectors

Attack vector	Attack types	Impact	Detection	Marked
Master station	Volumetrical/ logical	Data manipulation, DoS, reset	SM—Phase 2	1
Slave station	Volumetrical/ logical, reconnaissance	Data manipulation, DoS, reset	SM—Phase 1, 2	2
Trans. medium	MitM	Sniffing, traffic modification	SM—Phase 1	3

4 Modbus/TCP Security Module

In order to detect security incidents in industrial networks, the Security Module (SM) was designed and implemented in an experimental network. SM sniffs network traffic and parses Modbus/TCP industrial protocol traffic. To sniff network traffic, Scapy was used. The Scapy is integrated within Python script, the script process data parsed from the traffic medium, add data like *Time* and store them into the PostgreSQL database. The experimental testing uses Modbus/TCP protocol and its most common Function Codes. Table 3 summarize the Function Codes that are processed via SM. Each Function Codes define the parameters used in a request/response type of message. The transaction IDs, protocol IDs, lengths, and unit IDs are independent of the function codes and are thus always stored in the database. PostgreSQL database also stores data taken from Ethernet Layer, Internet Protocol layer, and Transmission Control Protocol layers.

An experimental network was created to implement security incidents detection using SM. The VMware software was used for virtualization of the Master and Slave station, see Fig. 2. The master and slave station run on Ubuntu operating system. The SM is implemented within each, master and slave station. For the implementation of the Modbus/TCP library, the Pymodbus[1] library was chosen and implemented. Normal communication via the Modbus/TCP protocol takes place through the "original" unsecured line (black marked), SQL (Structured Query Language) takes place through a dedicated and secured line to the parent element—SM master device. The SM master is used to evaluate and present the data obtained from the master and slave device databases. If the master generates a query: the first step is to capture Modbus communication using SM 1, the selected communication parameters (MAC, IP, TCP, Modbus/TCP layers, and timestamp produced by the SM with the accuracy of ms) are filtered. This information is added to existing records in the SM1 SQL database. Due to the need to work with large amounts of data, the PostgreSQL[2] database was chosen. The next step is to transmit the communication through an

---

[1] Pymodbus available on: https://pypi.org/project/pymodbus/.

[2] PostgreSQL available on: https://www.postgresql.org.

Table 3 Overview of selected Modbus/TCP parameters [4]

FC	Register type	Type	Specific for FC
1	Read coil	Request	startAddr, quantity
		Response	byteCount, coilstatus
2	Read discrete input	Request	startAddr, quantity
		Response	byteCount, inputStatus
3	Read holding register	Request	startAddr, quantity
		Response	byteCount, registerValue
5	Write single coil	Request	outputAddr, outputValue
		Response	outputAddrs, outputValue
6	Write single holding register	Request	registerAddr, registerValue
		Response	registerAddr, registerValue
15	Write multiple coils	Request	startAddr, quantity-Output,byteCount, outputValue
		Response	startAddr, quantityOutput

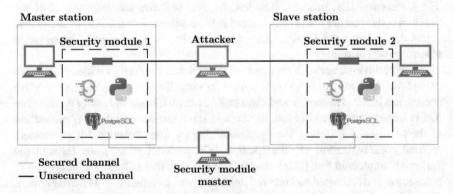

Fig. 2 Network setup, principle of security module [4]

unsecured channel to the slave side. The last step is to capture the communication using SM2 (identical steps are performed as in SM1). New records in the databases are gradually queried and evaluated by a two-Phase approach in the SM master device. The SM master establishes a secured connection using Secure Shell.

Flowchart of the Proposed Solution

The SM master first establishes a secure connection through SSH using pre-shared keys, see Fig. 3. During the first run of the SM master, the last record from the SM1 database is requested (Q_1). The SM1 database provides a response to this query, which is then processed (R_1) and the ID parameter is parsed and marked as *ID*. This ID (ID_{R1}) forms a counter to find the next record in the SM1 database. For the second and following runs is ID incremented to find another record, see Fig. 4, where N was set to 1 to get the maximal sensitivity and iterate over all records in the databases. The sequence number (SEQ_{R1}) is parsed from the obtained data, which is then used as a search element in the SM2 database (Q_2). The corresponding values of the individual records are compared with each other (Phase 1), if the content has not been altered or not found (MitM, injection attack), the record is predicted using machine learning or neural networks (Phase 2).

4.1   Phase 1—Databases Approach

The SM is able to evaluate (compare) the communication between SM1 and SM2 and on this basis to detect whether the message was generated from SM1, which was subsequently transmitted via the transmission medium to SM2 without any changes. During the evaluation, the individual parameters for cyber-attack detection are compared. The security alert is shown in Listing 1 (Modbus information is red marked, TCP blue marked, IP brown marked, MAC orange marked). This type of alert is generated every time, when the legitimate master station (SM 1) did not produce this message (another source—attacker generated this message), e.g. DoS, injection attack, or MAC spoofing.

Listing 1 Generated alert, no match found [4]

```
1  *** Security Violation ***
2  Record not found in master station!
3  [(15218, 1614757844594, 'request', 3, None, None, 10, 1, None,
      None, None, None, None, None, None, None, 3399, 0, 6,
      1, 45521, 502, 3242738942, 2527762142, 8, 0, 'PA', 502, 53482,
      0, None, 4, 5, 0, 64, 49522, 'DF', 0, 64, 6, 55023,
      '192.168.16.130', '192.168.16.131', '00:0c:29:32:a8:2a',
      '00:0c:29:0b:bc:20', 2048)]
```

If an attacker performs a MitM attack, the individual transmitted values are changed. In these cases, the security alert is generated, see Listing 2. Listing shows removal attack, where, the legitimate master station sent three function codes equal to three (red marked) but the legitimate slave station received only two of them (blue marked). If any value of the message is changed, a security alert is generated.

Listing 2 Generated alert, different parameters [4]

```
1  *** Security Violation ***
2  Problem detected in: Function Code.
3  Value of master is: [(3,), (3,), (3,)] value of slave is: [(3,),
      (3,)]
```

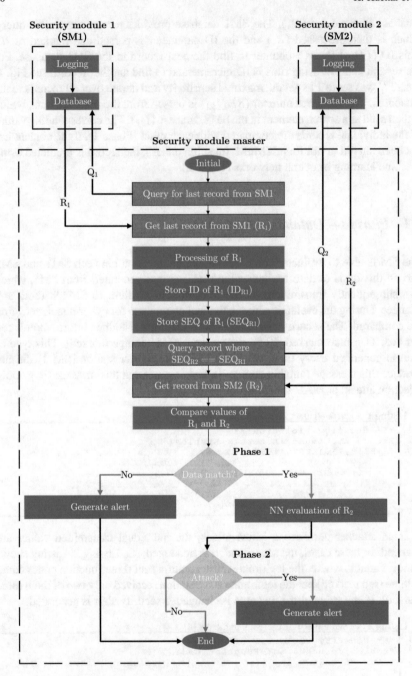

Fig. 3 The work flow of the first run [4]

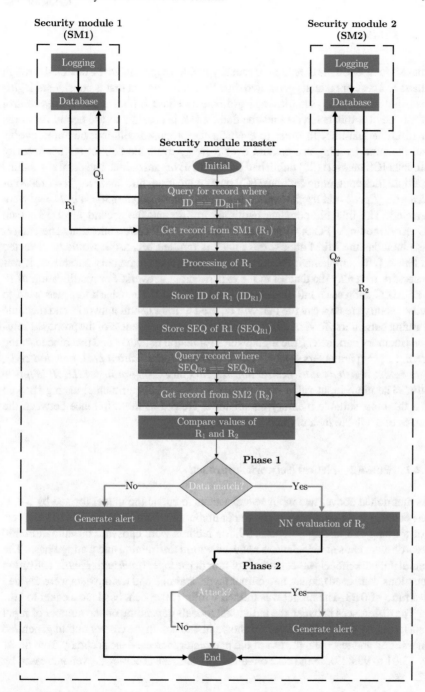

Fig. 4 The work flow of the second and following runs [4]

4.2   Phase 2

Phase 2 focuses on the detection of security incidents that cannot be detected through Phase 1. These are attacks generated directly from a station that is considered legitimate and is thus always marked as legitimate by Phase 1. If an attacker takes control of this device, this behavior must be detected. It is possible to use neural networks or machine learning. In order to evaluate the proposed solution, neural networks and machine learning are created over a freely available dataset (Cyber-security Modbus ICS dataset) [20] published by the IEEE organization. The dataset contains multiple files containing Modbus/TCP communication, files *clean traffic-eth2dump-clean-6h_1.pcap* and *traffic-eth2dump-modbusQueryFlooding30m-1h_1.pcap* were selected. The first file contains legitimate traffic and the second file DoS attack (Querry flooding—FC 3, 6) focused on Modbus/TCP communication. The dataset also includes the MitM attack, this attack is omitted because it would be detected in Phase 1. To supplement the dataset with the proposed memory parameter, it was necessary to replay the dataset in the experimental network. For modification of IP a nd MAC addresses, the Tcprewrite[3] tool was used. The tshark tool was used to subsequently create a csv file that will be used to train neural networks and machine learning classifiers. To verify the functionality and effectiveness of the proposed solution (memory parameter), the input values for neural networks and machine learning are 8, resp. 9 parameters + label. These are: *offset, destination port, function code, byte count, register value, transaction ID*, Modbus/TCP—*protocol ID, IP length + label*. The ninth input value is a *memory parameter*. Experimental testing showed that the most suitable memory parameter is created as the difference between the values of available memory and free memory.

4.2.1   Phase 2—Neural Network Approach

As mentioned above, two input sets of values to create the model formed by 8 + 1, respectively 9 + 1 parameters. The total number of records formed 100,101 records. For training and testing purposes, 99,501 records were chosen. For validation 600 records were chosen, which were not used within the training and testing Phase. The neural network model was consisting of four dense layer (*tanh* and *sigmoid* activation functions were used), input, and output layer. Training and testing data were divided in a ratio of 0.33. The model was trained within 150 epoch, batch size equal to 10.

The differences between the individual models depending on the number of input data (memory parameter usage) are shown in Table 4. In the case of output generated on training and test data, the use of the parameter increased the accuracy from 86.81 by 13.01 to 99.82%. In the case of validation data, the accuracy is even increased by 42.33%.

---

[3] Tcprewrite available on: https://tcpreplay.appneta.com/wiki/tcprewrite-man.html.

Table 4 Comparison of neural network models [4]

Phase	–		Model without Memory parameter	Model with memory parameter
Training/Testing	Total No. of data		99 501	99 501
	Parameters		8 + 1	9 + 1
	Training/Testing ratio		0.33	0.33
	No. of dense layers		4	4
	No. of epochs		150	150
	Loss		33.74%	2.82%
	Accuracy		86.81%	99.82%
Validation	Total No of data		600	600
	TP		76	300
	FP		30	0
	TN		270	300
	FN		224	0
	Precision		71.70%	100%
	Recall		25.33%	100%
	Accuracy		57.67%	100%

4.2.2 Phase 2—Machine Learning Approach

To demonstrate the ability to improve the accuracy of the predictive machine learning model against the approach when the memory parameter is not used we have chosen seven machine learning approaches and compared their outputs. The *Logistic Regression* (LR), *Random Forest, Linear Support Vector Classification* (LSVC), *K-Nearest Neighbours* (KNN), *Decision Tree, Gaussian Naïve Bayes* (GaussianNB), and *Support Vector Classification* (SVC). For the implementation of machine learning classifiers, the Scikit-learn[4] library was used. The overview of chosen machine learning classifiers is shown in Table 5.

The outputs were also predicted on the IEEE dataset with/without memory parameter. The dataset was divided into training, testing, and validation data. Validation data make up 20 % of the total dataset, training and test data make up 80 %. The distribution was performed randomly on pre-arranged data. Cross-validation was used to divide the data into training and testing, which are used during model training. Validation data were used to subsequently verify the already trained model, where labels were taken and records were passed to the model to make predictions (these data are not used to learn/train the model). Figure 5 graphically visualizes the differences from the accuracy point of view of the machine learning model with used memory parameter (orange line) and without this parameter (blue line). The grey line visualizes output accuracy based on the training/testing data (the orange line is the same for training/testing and validation data). If the memory parameter is used,

---

[4] Scikit-learn available on: https://scikit-learn.org/.

Table 5 Overview of selected machine learning clasifiers

Acronym	Machine learning clasifier
LR	LR(multi_class='ovr', solver='newton-cg')
	LR(multi_class='ovr')
	LR(multi_class='ovr', solver='liblinear')
	LR(multi_class='ovr', solver='sag')
	LR(multi_class='ovr', solver='saga')
	LR()
RFC	RFC()
LSVC	LSVC(dual=False, penalty='l1', random_state=0)
	LSVC(random_state=0)
	LSVC(loss='hinge', random_state=0)
KNN	KNN(n_neighbors=3)
	KNN()
	KNN(n_neighbors=7)
	KNN(n_neighbors=9)
	KNN(n_neighbors=11)
	KNN(weights='distance')
	KNN(algorithm='ball_tree')
	KNN(algorithm='kd_tree')
	KNN(algorithm='brute')
Decision tree	DecisionTree()
	DecisionTree(criterion='entropy')
	DecisionTree(splitter='random')
Gaussian NB	GaussianNB()
SVC	SVC(gamma='auto')
	SVC(kernel='linear')
	SVC(kernel='poly')
	SVC()
	SVC(kernel='sigmoid')

the accuracy is increased approximately by 0.19. In the case of SVC with the set kernel to *poly*, the machine learning model with memory parameter reaches the same accuracy as the model without memory parameter. All other models show improved accuracy. Table 7 show the outputs of selected machine learning classifiers without memory parameter. Table 8 shows the outputs when the memory parameter is used.

Figure 6 graphically visualizes the time required to perform training and testing (validation time is not included). The time needed for models without used memory parameter is blue marked and models with memory parameter are orange marked. It is clear from the figure that the use of the parameter reduces the learning time. Mutual comparison of TP, TN, FP, FN, accuracy, accuracy, precision, recall, and F1-score

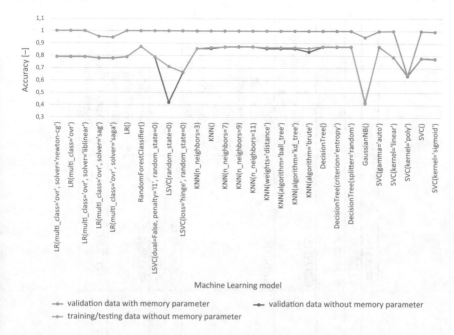

Fig. 5 Accuracy visualization of chosen machine learning classifier

is shown in Table 9. In the case of positive values, this is an advantage in favor of a machine learning model with memory parameter. In the case of a negative value, this is an advantage in favor of a machine learning model without memory parameter. In the case of the GaussianNB() model, the precision value is the same, so the value in the table is zero. The efficiency of using the memory parameter is obvious that a higher quality model is always achieved with one exception (SVC (kernel = 'poly')). The mathematical comparison is shown in Table 6. If the memory parameter is used • mark is used (◦ when memory parameter is not used). The table compares the machine learning models in the average, median, minimal, and maximal value from the point of accuracy and learning time view. If the memory parameter is used, the difference between the average accuracy values is higher by approx. 0.192. In the case of the median, the difference in accuracy is about 0.189. In terms of learning time, the use of the memory parameter reduces the average learning time by about 1 863.117 s. When using the median, the learning time differs by about 10.930 s. The largest difference in learning time was recorded when using SVC(kernel='linear') classifier.

Table 6 Statistical overview of the obtained accuracy values

		Without memory ○	With memory ●	Increase	Model
Accuracy (−)	Average	0.78928392	0.98137588	0.1920920	–
	Median	0.81092416	0.99969682	0.1887727	–
	Minimal value	0.42074335	0.63855483	0.2178115	○ LinearSVC(random_state=0) ● SVC(kernel='poly')
	Maximal value	0.87292818	1	0.1270718	○ RFC() ● DecisionTree() ● DecisionTree(criterion='entropy') ● DecisionTree(splitter='random') ● SVC(gamma='auto')
Time (s)	Average	02010.36852	0147.251143	−01863.1174	
	Median	00018.084403	0007.15472829	−00010.929675	
	Minimal value	00000.27593493	0000.15736246	−00000.1185725	○ GaussianNB() ● DecisionTreeClassifier(splitter='random')
	Maximal value	49813.731	2016.16422	−47797.567	○ SVC(kernel='linear') ● SVC(gamma='auto')

Table 7 Machine learning comparison—memory parameter not used

	Model setup	TP	TN	FP	FN	Accuracy	Precision	Recall	F1-score
Without memory parameter	LR(multi_class='ovr', solver='newton-cg')	11683	4052	3114	1061	0.790306379	0.789551936	0.916745135	0.848407828
	LR(multi_class='ovr')	11684	4048	3118	1060	0.790155701	0.789352790	0.916823603	0.848326436
	LR(multi_class='ovr', solver='liblinear')	11684	4048	3118	1060	0.790155701	0.789352790	0.916823603	0.848326436
	LR(multi_class='ovr', solver='sag')	11851	3658	3508	893	0.778955299	0.771600000	0.929927809	0.843397502
	LR(multi_class='ovr', solver='saga')	11851	3650	3516	893	0.778553491	0.771198022	0.929927809	0.843157483
	LR()	11684	4048	3118	1060	0.790155701	0.789352790	0.916823603	0.848326436
	RandomForestClassifier()	11054	6322	844	1690	0.872727273	0.929063708	0.867388575	0.897167438
	LSVC(dual=False, penalty='l1', random_state=0)	11684	4040	3126	1060	0.789753893	0.788926401	0.916823603	0.848080000
	LSVC(random_state=0)	1212	7165	1	11532	0.420743345	0.999175598	0.095103578	0.173676291
	LSVC(loss='hinge', random_state=0)	12617	615	6551	127	0.664590658	0.658232471	0.990034526	0.790737027
	KNeighborsClassifier(n_neighbors=3)	11145	5919	1247	1599	0.857056755	0.899370562	0.87452919	0.886775939
	KNeighborsClassifier()	11160	5929	1237	1584	0.858312406	0.900217795	0.875706215	0.887792848
	KNeighborsClassifier(n_neighbors=7)	10988	6358	808	1756	0.871220492	0.931502204	0.862209667	0.895517522
	KNeighborsClassifier(n_neighbors=9)	11028	6352	814	1716	0.872928177	0.931261611	0.865348399	0.897095908
	KNeighborsClassifier(n_neighbors=11)	10999	6376	790	1745	0.872677047	0.932988379	0.863072819	0.896669792
	KNeighborsClassifier(weights='distance')	11161	5923	1243	1583	0.858061276	0.89979039	0.875784683	0.887625258
	KNeighborsClassifier(algorithm='ball_tree')	11160	5933	1233	1584	0.85851331	0.900508351	0.875706215	0.887934121
	KNeighborsClassifier(algorithm='kd_tree')	11160	5929	1237	1584	0.858312406	0.900217795	0.875706215	0.887792848
	KNeighborsClassifier(algorithm='brute')	11436	5120	2046	1308	0.831541939	0.848242101	0.897363465	0.872111645
	DecisionTreeClassifier()	11049	6326	840	1695	0.872677047	0.929346455	0.866928207	0.897052854
	DecisionTreeClassifier(criterion='entropy')	11048	6326	840	1696	0.872626821	0.929340511	0.866917765	0.897044495
	DecisionTreeClassifier(splitter='random')	11049	6327	839	1695	0.872727273	0.929424630	0.866928207	0.89708927
	GaussianNB()	1212	7166	0	11532	0.420793571	1.000000000	0.095103578	0.173688736
	SVC(gamma='auto')	11043	6336	830	1701	0.872877951	0.930093489	0.866525424	0.897184872
	SVC(kernel='linear')	11703	4005	3161	1041	0.788950276	0.787338536	0.918314501	0.84779774
	SVC(kernel='poly')	12719	14	7152	25	0.639527875	0.640078506	0.998038293	0.779947877
	SVC()	11800	3727	3439	944	0.779859367	0.774329024	0.925925926	0.843369188
	SVC(kernel='sigmoid')	11815	3619	3547	929	0.775188348	0.769105585	0.92710295	0.840745748

Table 8 Machine learning comparison—memory parameter used

	Model setup	TP	TN	FP	FN	Accuracy	Precision	Recall	F1-score
With memory parameter	LR(multi_class='ovr', solver='newton-cg')	12640	7149	1	0	0.999949469	0.999920892	1.000000000	0.999960000
	LR(multi_class='ovr')	12640	7149	1	0	0.999949469	0.999920892	1.000000000	0.999960000
	LR(multi_class='ovr', solver='liblinear')	12640	7149	1	0	0.999949469	0.999920892	1.000000000	0.999960000
	LR(multi_class='ovr', solver='sag')	12434	6441	709	206	0.953764528	0.946054934	0.983702532	0.964511500
	LR(multi_class='ovr', solver='saga')	12312	6449	701	328	0.948004042	0.946130792	0.974050633	0.959887732
	LR()	12640	7149	1	0	0.999949469	0.999920892	1.000000000	0.999960000
	RandomForestClassifier()	12640	7149	1	0	0.999949469	0.999920892	1.000000000	0.999960000
	LSVC(dual=False, penalty='l1', random_state=0)	12640	7149	1	0	0.999949469	0.999920892	1.000000000	0.999960000
	LSVC(random_state=0)	12640	7149	1	0	0.999949469	0.999920892	1.000000000	0.999960000
	LSVC(loss='hinge', random_state=0)	12640	7149	1	0	0.999949469	0.999920892	1.000000000	0.999960000
	KNeighborsClassifier(n_neighbors=3)	12639	7145	5	1	0.999696817	0.999604556	0.999920886	0.999762696
	KNeighborsClassifier()	12640	7144	6	0	0.999696817	0.999525542	1.000000000	0.999762715
	KNeighborsClassifier(n_neighbors=7)	12640	7144	6	0	0.999696817	0.999525542	1.000000000	0.999762715
	KNeighborsClassifier(n_neighbors=9)	12640	7144	6	0	0.999696817	0.999525542	1.000000000	0.999762715
	KNeighborsClassifier(n_neighbors=11)	12640	7144	6	0	0.999696817	0.999525542	1.000000000	0.999762715
	KNeighborsClassifier(weights='distance')	12639	7145	5	1	0.999696817	0.999604556	0.999920886	0.999762696
	KNeighborsClassifier(algorithm='ball_tree')	12640	7144	6	0	0.999696817	0.999525542	1.000000000	0.999762715
	KNeighborsClassifier(algorithm='kd_tree')	12640	7144	6	0	0.999696817	0.999525542	1.000000000	0.999762715
	KNeighborsClassifier(algorithm='brute')	12640	7144	6	0	0.999696817	0.999525542	1.000000000	0.999762715
	DecisionTreeClassifier()	12640	7150	0	0	1.000000000	1.000000000	1.000000000	1.000000000
	DecisionTreeClassifier(criterion='entropy')	12640	7150	0	0	1.000000000	1.000000000	1.000000000	1.000000000
	DecisionTreeClassifier(splitter='random')	12640	7150	0	0	1.000000000	1.000000000	1.000000000	1.000000000
	GaussianNB()	11638	7150	0	1002	0.949368368	1.000000000	0.920727848	0.958728067
	SVC(gamma='auto')	12640	7120	30	0	0.998484083	0.997632202	1.000000000	0.998814698
	SVC(kernel='linear')	12640	7150	0	0	1.000000000	1.000000000	1.000000000	1.000000000
	SVC(kernel='poly')	12625	12	7138	15	0.638554826	0.638820017	0.998813291	0.779248835
	SVC()	12629	7148	2	11	0.999343103	0.999841659	0.999129747	0.999485576
	SVC(kernel='sigmoid')	12534	7140	10	106	0.994138454	0.999202806	0.991613924	0.995393901

Table 9 Difference of machine learning classifiers

Model	Difference			
	Accuracy	Precision	Recall	F1-score
LR(multi_class='ovr', solver='newton-cg')	0.209643091	0.210368956	0.083254865	0.151552616
LR(multi_class='ovr')	0.209793769	0.210568102	0.083176397	0.151634009
LR(multi_class='ovr', solver='liblinear')	0.209793769	0.210568102	0.083176397	0.151634009
LR(multi_class='ovr', solver='sag')	0.174809229	0.174455221	0.053774722	0.121113998
LR(multi_class='ovr', solver='saga')	0.169450552	0.174932771	0.044122824	0.11673025
LR()	0.209793769	0.210568102	0.083176397	0.151634009
RFC()	0.127222197	0.070857184	0.132611425	0.102793007
LSVC(dual=False, penalty='l1', random_state=0)	0.210195577	0.210994491	0.083176397	0.151880311
LSVC(random_state=0)	0.579206124	0.000745295	0.904896422	0.826284153
LSVC(loss='hinge', random_state=0)	0.335358811	0.341688422	0.009965474	0.209223418
KNN(n_neighbors=3)	0.142640061	0.100233994	0.125391696	0.112986757
KNN()	0.141384411	0.099307747	0.124293785	0.111969866
KNN(n_neighbors=7)	0.128476324	0.068023338	0.137790333	0.104245192
KNN(n_neighbors=9)	0.12676864	0.06826393	0.134651601	0.102666806
KNN(n_neighbors=11)	0.12701977	0.066537163	0.136927181	0.103092923
KNN(weights='distance')	0.141635541	0.099814165	0.124136203	0.112137437
KNN(algorithm='ball_tree')	0.141183507	0.09901719	0.124293785	0.111828594
KNN(algorithm='kd_tree')	0.141384411	0.099307747	0.124293785	0.111969866
KNN(algorithm='brute')	0.168154878	0.151283441	0.102636535	0.127651070
DecisionTree()	0.127322953	0.070653545	0.133071793	0.102947146
DecisionTree(criterion='entropy')	0.127373179	0.070659489	0.133082235	0.102955505
DecisionTree(splitter='random')	0.127272727	0.070575370	0.133071793	0.10291073
GaussianNB()	0.528574797	0.000000000	0.82562427	0.785039331
SVC(gamma='auto')	0.125606132	0.067538713	0.133474576	0.101629826
SVC(kernel='linear')	0.211049724	0.212661464	0.081685499	0.152202260
SVC(kernel='poly')	-0.00097305	-0.001258489	0.000774999	-0.000699042
SVC()	0.219483735	0.225512635	0.073203821	0.156116388
SVC(kernel='sigmoid')	0.218950106	0.230097221	0.064510974	0.154648153

5 Graphical Visualization

The Security Modules use databases as mentioned above. In our approach we have chosen open-source analytical and monitoring tool Grafana[5] to visualize the content of sniffed are stored records. This tool periodically queries new content in databases and represents taken data in a graphical way (tables, graphs). The graphical representation enables to detection of changes in network traffic. Also, it is possible for this software as a human-machine interface for visualization of chosen data (coils/registers) to provide easily understandable process visualization. To be able to detect changes in transmitted data (coil/register values), it is crucial to handle with time precisely (done when logging data into the database—Security Module). When time

---

[5] Grafana available on: https://grafana.com/.

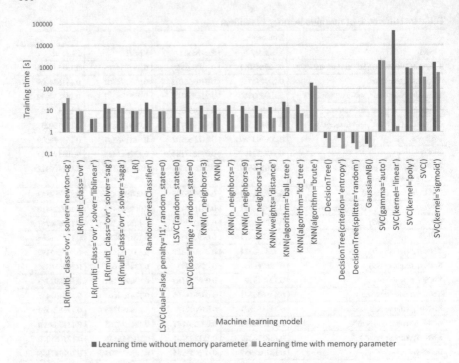

Fig. 6 Learning time visualization of chosen machine learning classifier

is stored with an accuracy of seconds, graphs are redrawn and misinterpreted. It is thus necessary to pass the time with an accuracy of milliseconds. The *registerval* parameter takes different values, therefore it is stored in the Security Module in text format, for visualization, it is necessary to perform casting to the value integer.

6 Conclusion

This paper focused on incident detection techniques in an industrial network, where the secured industrial protocol is not placed. This additional security is reached using so-called Security Modules. For the implementation of the security module, no network adjustments are needed. These modules sniff Modbus/TCP traffic, process it, and store it into the local PostgreSQL database. The parent Security Module master uses the data stored in the databases in Security Module 1 and 2 for incident detection. The distinction between legitimate data traffic and attack is done in two Phases. Phase 1 uses databases comparison to detect e.g. traffic generated from non-master station, MitM attack, command injection, in other words, all data traffic that is not originating from the legitimate master station. This approach is supposed to secure a legitimate station, for cases when the master station is controlled by the attacker, Phase 2 was developed. Phase 2 uses a neural network or machine learning to detect infected master station. To provide a reliable result, the publicly available IEEE dataset was used for training and testing. The Security Module also uses memory parameter taken from the slave station. Based on the experimental testing, the memory parameter significantly improves the accuracy of the trained model, both neural network and machine learning.

The main benefit of this paper is machine learning classifiers comparison. The seven machine learning classifiers, where the setting parameter of each classifier was adjusted to provide a wide comparison of available predictive machine learning classifiers comparison. In total 28 predictive machine learning models were trained and tested with the following validation. As the results in the experimental testing showed, the usage of memory parameter increases the accuracy of the machine learning model approximately by 20% (average value based on validation data). The time needed for training is also decreased by approximately about 10.93 s (median value based on the validation data). Using the mentioned approach we are able to get accuracy in the range 0.64–1 (when memory parameter is used) or range 0.42–0.87 (without memory parameter). As the best approach seems to be using *Decision Tree* or *Support Vector Classification* (with set gamma to "auto") classifiers (accuracy equal to 1). The worst results have been achieved *Support Vector Classification* (with the set kernel to "poly") (accuracy equal to 0.64). As future work, we would like to focus on the wider area of attack, especially logic attacks, implement the presented solution into the docker and make this approach protocol independent. Even there is no additional delay in the communication, the query parameter that searches the records in the Security Module 2 database takes a bit of time in case of too frequent communication, this behavior could be solved by running the scripts in multi-threading processing. The proposed solution provides many features and could be implemented also into the already secured network to provide an even higher perspective over the network and data traffic.

Acknowledgements The described research is part of the grant project registered under no. VI20192022132 and funded by the Ministry of Interior of the Czech Republic.

References

1. Stouffer K, Pillitteri V, Lightman S, Abrams M, Hahn A (2015) National institute of standards and technology special publication 800–82, revision 2. U.S. Department of Commerce, NIST
2. (2020) Security for industrial automation and control systems, 1st edn.
3. North American electric reliability corporation (2008). Available: https://www.nerc.com/pa/Stand/Pages/CIPStandards.aspx
4. Holasova E, Kuchar K, Fujdiak R, Blazek P, Misurec J (2022) Security modules for securing industrial networks. In: International conference on information system and network security (CISNS 2021), vol 2, pp 1–8

5. Beaver JM, Borges-Hink RC, Buckner MA (2013) An evaluation of machine learning methods to detect malicious SCADA communications. In: 2013 12th International conference on machine learning and applications, pp 54–59. Available: http://ieeexplore.ieee.org/document/6786081/

6. Mubarak S, Habaebi MH, Islam MR, Khan S (2021) ICS cyber attack detection with ensemble machine learning and dpi using cyber-kit datasets. In: 2021 8th International conference on computer and communication engineering (ICCCE), pp 349–354. Available: https://ieeexplore.ieee.org/document/9467162/

7. Joshi C, Khochare J, Rathod J, Kazi F (2020) A semi-supervised approach for detection of SCADA attacks in gas pipeline control systems. In: 2020 IEEE-HYDCON, pp 1–8. Available: https://ieeexplore.ieee.org/document/9242676/

8. Alhaidari FA, AL-Dahasi EM (2019) New approach to determine DDoS attack patterns on SCADA system using machine learning. In: 2019 International conference on computer and information sciences (ICCIS), pp 1–6. Available: https://ieeexplore.ieee.org/document/8716432/

9. Bulle BB, Santin AO, Viegas EK, dos Santos RR (2020) A host-based intrusion detection model based on OS diversity for SCADA. In: IECON 2020 The 46th annual conference of the IEEE industrial electronics society, pp 691–696. Available: https://ieeexplore.ieee.org/document/9255062/

10. Perez RL, Adamsky F, Soua R, Engel T (2018) Machine learning for reliable network attack detection in SCADA systems. In: 2018 17th IEEE international conference on trust, security and privacy in computing and communications/12th IEEE international conference on big data science and engineering (TrustCom/BigDataSE), pp 633–638. Available: https://ieeexplore.ieee.org/document/8455962/

11. Knapp E (2011) Chapter 4—Industrial network protocols. In: Industrial network security. In: Knapp E (ed) Syngress, Boston, pp 55–87. Available: https://www.sciencedirect.com/science/article/pii/B9781597496452000045

12. Chang H-C, Lin C-Y, Liao D-J, Koo T-M (2020) The modbus protocol vulnerability test in industrial control systems. In: International conference on cyber-enabled distributed computing and knowledge discovery (CyberC), pp 375–378

13. Yue G (2020) Design of intelligent monitoring and control system based on modbus. In: 2020 5th International conference on communication, image and signal processing (CCISP), pp 149–153

14. Radoglou-Grammatikis P, Siniosoglou I, Liatifis T, Kourouniadis A, Rompolos K, Sarigiannidis P (2020) Implementation and detection of modbus cyberattacks. In: 2020 9th International conference on modern circuits and systems technologies (MOCAST) pp 1–4

15. Zhou X, Xu Z, Wang L, Chen K, Chen C, Zhang W (2018) Kill chain for industrial control system. In: MATEC web conference, vol 173. Available: https://doi.org/10.1051/matecconf/201817301013

16. Fachkha C (2019) Cyber threat investigation of SCADA modbus activities. In: 2019 10th IFIP international conference on new technologies, mobility and security (NTMS), pp 1–7

17. Nardone R, Rodríguez RJ, Marrone S (2016) Formal security assessment of modbus protocol. In: 2016 11th International conference for internet technology and secured transactions (ICITST), pp 142–147

18. Luswata J, Zavarsky P, Swar B, Zvabva D (2018) Analysis of SCADA security using penetration testing: a case study on modbus TCP protocol. In: 2018 29th Biennial symposium on communications (BSC), pp 1–5

19. Volkova A, Niedermeier M, Basmadjian R, de Meer H (2019) Security challenges in control network protocols: a survey. IEEE Commun Surveys Tutor 21(1):619–639

20. Frazão I, Abreu P, Cruz T, Araújo H, Simões P (2019) Cyber-security modbus ICS dataset. Available: https://dx.doi.org/10.21227/pjff-1a03

Predictive Maintenance of Vehicle Fleets Using LSTM Autoencoders for Industrial IoT Datasets

Arindam Chaudhuri, Rajesh Patil, and Soumya K. Ghosh

Abstract Connected vehicle fleets have often formed significant component of industrial internet of things scenarios worldwide in lines with Industry 4.0 standards. The number of vehicles in these fleets have grown steadily. The monitoaring of these vehicles with machine learning algorithms has significantly improved maintenance activities of these systems. Since past few decades predictive maintenance potential has increased with machines being controlled and managed through networked smart devices. Predictive maintenance has provided benefits with respect to uptimes optimization. This has resulted in reduction of time and labor costs associated with inspections and preventive maintenance. It has also provided significant cost benefit ratios in terms of business profits. In order to appreciate vehicle fault trends considering important vehicle attributes, this problem is addressed through LSTM Autoencoders. This acts as predictive analytics engine for this problem. Real world data is collected from vehicle garages are used for training and testing of LSTM Autoencoders. This method is compared with several support vector machine variants. This process helps better implementation of telematics data in order to ensure preventative management towards desired solution. The superiority of this method is highlighted through several experimental results.

Keywords Predictive maintenance · Industrial IoT · LSTM autoencoders · Vehicle fleets · RUL

A. Chaudhuri (✉)
Samsung R & D Institute Delhi, Noida 201304, India
e-mail: arindam.chaudhuri@nmims.edu

A. Chaudhuri · R. Patil
NMIMS University, Mumbai 400056, India

S. K. Ghosh
Department of Computer Science Engineering, Indian Institute of Technology Kharagpur, Kharagpur 701302, India

© Springer Nature Switzerland AG 2022
R. Jiang et al. (eds.), *Big Data Privacy and Security in Smart Cities*,
Advanced Sciences and Technologies for Security Applications,
https://doi.org/10.1007/978-3-031-04424-3_6

1 Introduction

The connected vehicles [1] in present business setup forms an integral part of operations in every industry today. These systems have evolved with sensor-based data over the years. The ever-increasing digitization of modern vehicle systems has allowed generation of huge volumes of digitized data. The data analysis reveals underlying patterns which are not visible to human beings and supports proactive decision making. Organizations have struggled to capture and harness power of internet of things (IoT) information [2]. This information can readily be applied for operational insights considering devices and get ahead of unplanned downtime. Organizations always look for faster ways [3] in order to realize sensor information and transform it into predictive maintenance insights [4].

The vehicle downtime cost plays significant role towards demand of customers for higher up-times with aggressive service level agreements. The service providers look for predictive maintenance methods using accurate real-time vehicle information. This helps them in determining vehicle's condition and its required maintenance. This approach provides savings [5] with respect to cost. The predictive maintenance allows scheduling for corrective maintenance with unexpected vehicle failures prevention. The maintenance work can always be better planned considering apriori information. With connected vehicles predictive maintenance solution [6] users achieve various benefits. These include timely maintenance towards increased up-times, better plan maintenance in reducing unnecessary field service calls, optimizing repair parts replacement, reducing unplanned stops, improving vehicle performance and service compliance reporting. Vehicles are designed with temperature, infrared, acoustic, vibration, battery-level and sound sensors in order to monitor conditions which form as initial maintenance indicators as shown in Fig. 1. The predictive maintenance

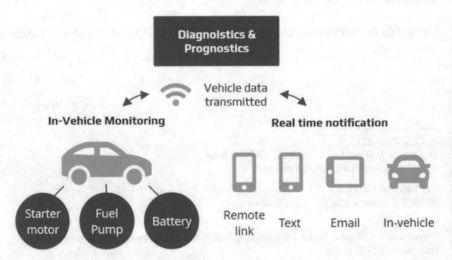

Fig. 1 Predictive maintenance of vehicles

programs are driven by customers which helps them to collect and manage vehicle data alongwith visualization and analytics [7] tools to make better business decisions.

Organizations have sent field technicians towards regular diagnostic inspections and preventive maintenance [8, 9] as per prespecified schedule of vehicles. Here an appreciable amount of cost and labor is involved with little assurance that failure would not happen in inspections. Despite traditional preventive measures put in place to avoid vehicle downtime, asset maintenance is often rife with unexpected failure which requires unplanned maintenance. Traditional checklists or quality control procedures provide insight but only after the vehicles have begun to show signs of failure or have already failed.

Preventive maintenance is required by most organizations [8, 9] but with schedules that have been predetermined by vehicle manufacturers. These schedules often lead to replacement of repair parts well before they are required. Unnecessary part replacement and unrequired servicing creates tremendous expenses especially considering the fact that maintenance may not be based on any operational data preventive analysis. When unforeseen vehicle failures occur organizations typically move to reactive maintenance model. This is prohibitively expensive and substantially inefficient. Due to lack of information service technicians are left scrambling to limit operational losses and bring vehicle back online. Predictive maintenance using advanced analytics catches the issues early and at source to identify and address potential vehicle issues and failures before they happen. It also optimizes end-to-end service operations and results in an overall improvement in vehicle reliability and performance.

Closely related to maintenance of connected vehicles is smart vehicle systems where Industry 4.0 comes in picture in manufacturing domain where optimization is performed through water and steam power, mass production in assembly lines and automation using information technology. Industry 4.0 presents an overall change by digitalization and automation of every part of an organization. Big international organization use concepts of continuous improvement and have high standards for research and development which accept concepts of Industry 4.0 and make themselves more competitive in market. This becomes possible by introducing self-optimization, self-cognition and self-customization into industry.

In order to solve problems posed by predictive maintenance of vehicle fleets in specified garages for telecom-based company, giving due consideration to Industry 4.0 standards here deep learning model is used to develop real time solutions. The motivation behind this work is adopted from success obtained by using support vector machine (SVM) variants such as hierarchical modified fuzzy support vector machine (HMFSVM), modified fuzzy support vector machine (MFSVM) and SVM [8, 9] and [10] in order to address similar problem. In this work LSTM Autoencoder is used to achieve predictive analytics task. It helps in determining health condition of vehicle's machine. The autoencoders allow learning compressed representation of input data and readily change input data dimensions. LSTM processes time series data and identifies temporal data patterns. Here LSTM Autoencoder comprises of an autoencoder for sequential data combined with LSTM. The historical time series data is trained by LSTM encoder. The records from previous maintenance activities produce fixed length vector which is imported to decoder. This is being classified as one of the predetermined labels. After classification is achieved, its output is associated with remaining useful time (RUL). This allows predictive maintenance actions to happen. This method is compared with SVM variants. The experimental results support LSTM Autoencoder's superiority over other algorithms.

This paper is organized as follows. In Sect. 2 related work is presented. In Sect. 3 computational method of LSTM Autoencoder is highlighted. This is followed by experiments and results in Sect. 4. Finally in Sect. 5 conclusions are given.

2   Related Work

During past decade, condition monitoring of vehicle machines has attracted high interest among researchers [11–14]. With continuous progress in artificial intelligence and associated disciplines, predictive maintenance has evolved towards identification of abnormalities in large scale datasets. This calls for determination of imminent maintenance needs [15]. The adoption of artificial intelligence and in particular machine learning methods for predictive maintenance prevents machine failures without requiring a clear understanding of production process [16, 17]. The reason is that data driven approaches are used in order to train machine learning models with run-to-failure data without any knowledge of underlying process [17, 18].

Several machine learning methods have been investigated throughout the years. An overview of machine learning approaches for predictive maintenance is presented in [19] and [20]. Artificial neural networks combined with data mining tools [21] and bayesian networks [22] have been used towards large manufacturing datasets to diagnose and predict faults. This presents issues associated with process time and computational learning due to large amount of data. For sequence-to-sequence learning, transformer models have recently received increased attention as highlighted in [23] and [24].

Convolutional neural networks (CNNs) have been suggested for fault diagnosis on multi-channel data from sensors with excellent performance and lower computational cost [25]. Double deep autoencoder proposed by [26] for clustering distributed and heterogeneous datasets addresses this problem. Autoencoders consist of exactly one input and output layer and one or more hidden layers. The input values are compressed at encoding stage and then same values are reconstructed at decoding stage [27]. For sequential data such as time-series data, recurrent neural networks (RNNs) are considered more suitable than CNNs [28, 29]. RNNs contain feedback loops and remember information from former units. Although they are capable of capturing long term temporal dependencies from data, they have restrictions on long term RUL predictions [30]. Small gradients tend to slowly shrink and eventually disappear during propagation across multiple unfoldings of network layers [31]. Popular variants of RNNs that avoid these limitations are long short-term memory (LSTM) networks and gated recurrent unit (GRU) networks [29, 32–34]. LSTM hidden structure includes new unit, memory cell capable of representing long-term dependencies in sequential time series data contrary to GRU networks [35–38]. Nevertheless, because of sensitivity to dataset changes or program tuning parameters, internal layers and cells in LSTM networks seem to lack efficiency [39].

LSTM memory cells consist of different units commonly known as gates. Gates control interactions between memory units and decide what portion of data is relevant to keep or forget during training. The input gate is responsible to decide if memory cell state can be modified by input signal. The output gate determines if other memory cells states can be modified by input signal. Finally, forget gate is in charge of deciding whether to forget or remember previous signal status [40, 41]. LSTM Autoencoders are capable of dealing with data sequences as input in contrast to regular autoencoders. Many studies have presented LSTM Autoencoders as promising tool for time series predictions. LSTM Autoencoder has been used for traffic flow prediction and has shown an outstanding performance in mining deeply big data considering temporal characteristics. It also captures spatial characteristics of traffic flow in comparison with CNN and SVM models [42]. Stacked autoencoders with LSTM have been used to predict one-step-ahead closing price of six popular stock indices traded in different financial markets with superior results [43]. Autoencoders combined with LSTM have been presented as best performing model in terms of RMSE values for different training and test datasets. This shows better capability of feature extraction of these models allowing better forecast than multilayer perceptron, deep belief network and single LSTM [44]. The combination of autoencoders and LSTM have shown high potential in time series prediction.

The purpose of deep learning LSTM Autoencoder network is to gather and extract composite information from large time-series datasets using many hidden layers [45]. However, choosing suitable hyperparameters is complex task and significantly affects model's performance [46]. For example, more hidden layers or more neurons in network does not necessarily increase network's performance. Hyperparameter selection depends on different factors such as the amount of data or generating process [45].

3 Computational Method

In this section framework of LSTM Autoencoder model is presented. The research problem entails in predictive maintenance for vehicle fleets which in a way involves certain preventive maintenance [47, 48] activities. In order to achieve this, we propose deep learning-based predictor viz LSTM Autoencoder to analyze vehicle fleets' maintenance. The prediction task on various strategic aspects of vehicle analytics provides information for several decision-making activities.

3.1 Datasets

The datasets are adopted from telecom-based company [9] from a specified garage. The datasets available for predictive analytics include data from 5 garages. The data from the first garage spanned over a period of 14 months and 24 months for other garages. The number of data records available from first garage was 3923 of which 2097 was used for prediction analysis. The number of data records available from other garages was 890,665 of which 11,456 was used for prediction analysis. The reasons behind not using the entire available data for prediction include:

(a) there are data rows with invalid values (null, value out of range, blanks etc.) which have been considered as outliers and have been removed
(b) vehicle repair types count (audits, automatic transmission, body exteriors, body interiors etc.) did not have considerable values for all vehicles are removed
(c) fuel economy score and driver behavior score were available only for first garage which was not used because it reduces prediction accuracy.

There was great degree of skewedness in the training data. Before applying the data to the prediction engine, it was balanced through external imbalance learning method. The dataset was used after the majority class under-sampling which makes the number of good training samples equal to the rejected training examples.

3.2 LSTM Autoencoder for Predictive Maintenance

A supervised deep learning-based approach is used in order to predict health of vehicle. Sensor data which monitors different parameters of vehicle are captured. They are used towards training LSTM Autoencoders set. Then trained LSTM Autoencoders classifies new streams of incoming data with respect to different operational status. A schematic representation of this system is highlighted in Fig. 2.

After classification an estimation of RUL value is determined based on dataset's performance. This depends on time which vehicle's machine required for issuing a replacement order in past providing similar values to input. The datasets are placed on vehicle's machine measuring key process-related features from equipment and its environment. On processing this generates required features for enabling learning process. Nevertheless, it is challenging to identify correct set of features which are associated towards potential failures. The critical values are determined and modeled considering vehicle's machine degradation process. Hence, key features used for analysis are selected using knowledge stemming from vehicle's machine related studies. LSTM Autoencoders are used for classifying current vehicle's health condition to one or more corresponding labels. The architecture of each LSTM-autoencoder depends on problem's nature [49, 50]. Basically, two labels are required for determining good condition of vehicle's equipment. This refers towards:

(a) a time right after maintenance activity or part replacement has happened

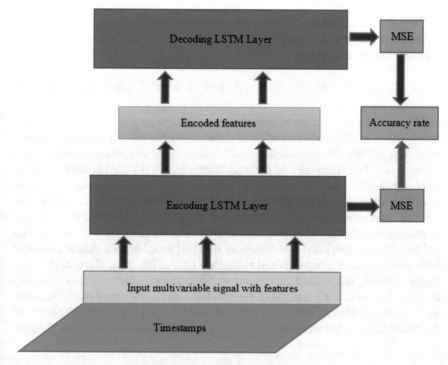

Fig. 2 Architecture of LSTM autoencoder for predictive maintenance

(b) a label characterizing a failure, an alarm or a low health-level from an operational perspective calling for maintenance to be conducted.

In this work, three labels are considered in order to determine high, medium and low level of vehicle's health status. The high and low levels are good and bad labels corresponding to an observation window right before and after restoration activity. This brings vehicle's machine to its normal operational condition. This also include vehicle's machine breakdown. With each label a single LSTM-autoencoder is trained such that training dataset includes only temporal data related with corresponding vehicle's machine status. Each LSTM Autoencoder inputs a time-series sequence which is denoted as B_i with β_{i_j} having values of sensors denoting one of the variables measured at specific time with n being number of features:

$$B_i = [\beta_{i_1}, \beta_{i_2}, \beta_{i_3}, \beta_{i_4}, \beta_{i_5}, \ldots, \beta_{i_j}] \forall \beta_{i_j} \in \mathbb{R} \text{ with } i, j \in \mathbb{Z} \text{ and } i \leq n \quad (1)$$

Each time-series sequence is placed into new encoder LSTM cell together with hidden output from previous LSTM cell. The hidden output from encoder's last LSTM cell is encoded finally into a representation vector. This vector can be considered as hidden states sequence from all previous encoder LSTM cells. Then decoder takes encoded features as input in order to be processed through various

LSTM decoder cells which finally produces output. The decoder layer's output is a reconstruction of initial input time-series sequence which is represented as B_i' with reconstructed sensor values as:

$$B_i' = \left[\beta_{i_1}', \beta_{i_2}', \beta_{i_3}', \beta_{i_4}', \beta_{i_5}', \ldots, \beta_{i_j}' \right] \forall \beta_{i_j}' \in \mathbb{R} \text{ with } i, j \in \mathbb{Z} \text{ and } i \leq n \quad (2)$$

This requires segregation of sensor data in order to properly train each autoencoder. This is made possible by using actual maintenance records and expert knowledge. The previous maintenance records contain dates and causes which require maintenance. This is provided by preventive maintenance plans set in place by equipment owner. When this information is combined with historical sensor data values from these specific dates, it is possible to label data values and distinguish datasets for network training and testing. Data values are categorized and segregated according to number and chosen status considering their timestamp. In order to define, train and test data for each LSTM Autoencoder split performed in each dataset considers 90% of dataset as training data and remaining 10% as test data. The accuracy of a model is usually determined after model training. Test samples are fed as input into model and network compares initial input values with reconstructed ones. The mean squared error (MSE) of difference between reconstructed time series sequence B_i' and initial input B_i is cost function of LSTM Autoencoder which is represented as:

$$MSE_i = \frac{1}{n} \sum_{i=1}^{n} \left(B_i' - B_i \right)^2 \quad (3)$$

For each time-sequence a mean squared error is calculated. Then, accuracy rate of LSTM Autoencoder is obtained from average calculation of these values. After training of LSTM Autoencoders set, same dataset becomes input to each of these trained networks. Then, according to accuracy rate which appears with this input, classification of input dataset is performed. Since three labels have been used here in order to determine high, medium and low level of vehicle's machine health status, we use three LSTM Autoencoders. These are trained and each of them with a dataset includes only temporal data related with corresponding vehicle's machine status.

The stated approach has been implemented using Python 3.7 with Intel i7 processor on Microsoft Windows 10. LSTM Autoencoder parameters are trained with more than 300,000 data points. For each LSTM Autoencoder RUL prediction results are obtained in appreciable times. A schematic representation of system's architecture from implementation perspective is presented in Fig. 3.

The architecture of each LSTM Autoencoder included an input layer where size depends on number of features selected. Here, 7 features are considered. The first encoding LSTM layer reads input data and outputs sixteen 49 features with one timestep for each. The second encoding LSTM layer reads 1×49 inputs from first encoding LSTM layer and reduced feature size to 7. The output of this layer is 1×7 feature vector. A repeat vector replicated feature vector and prepared 2D array input for first LSTM layer in decoder. Then, first decoding LSTM layer read 1×7 input

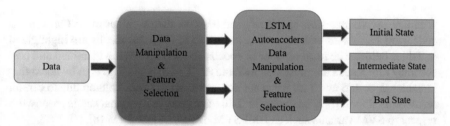

Fig. 3 Schematic representation of system's architecture from implementation perspective

data and outputs 7 features with one timestep for each. The second decoding LSTM layer reads 1×7 inputs and outputs 49 features with one timestep for each. Then, time distributed layer took output and created 49×7 vector. This corresponds to number of features output from previous layer multiplied by number of features. Finally, a matrix multiplication between second decoding LSTM layer and time distributed layer outputs 1×7 output.

4 Experiments and Results

The performance evaluation is performed considering data from 12 months of vehicle operation. The datasets are created considering historical maintenance records of vehicle's machine states. The architecture of LSTM Autoencoder including network layers, number of parameters with weights and biases of each layer and total model parameters are highlighted in Table 1.

The network's processing complexity is affected by parameters [51]. The total number of parameters trained in each LSTM Autoencoder network is 3750. The

Table 1 LSTM autoencoder parameters

Layer	Network type	Output shape	Parameters
input-1	Input layer	1×7	0
lstm-1	LSTM	1×49	1494
lstm-2	LSTM	1×7	336
Repeat vector-1	Repeat vector	1×7	0
lstm-3	LSTM	1×7	300
lstm-4	LSTM	1×49	1494
timedistributed-1	Time distributed	1×7	126
Total parameters			3750
Trainable parameters			3750
Non-trainable parameters			0

accuracy, recall, precision, specificity and F1 score metric values of LSTM Autoencoders are calculated with respect to test data [52, 53]. All results are highlighted in Tables below. The ideal number of epochs and batch sizes are identified and best accuracy rate results are presented in Tables 2, 3, 4, 5, 6 and 7. It is to be noted that in Tables 2, 3, 4, 5 and 7 all dates are not presented in Dates column due to certain restrictions imposed by business. In Table 6 a comparative analysis is made with respect to SVM variants such as HMFSVM, MFSVM and SVM [8].

Each LSTM Autoencoder is trained with number of epochs and batch sizes presented by accuracy rate. The initial state, intermediate state and final state LSTM Autoencoder presented appreciable accuracy results after training with 50, 70 and 100 epochs respectively. For each timestamp, in Table 8 first two columns represent actual states of monitored vehicle's machine at specific timestamps while next to them is provided accuracy rate generated by each LSTM Autoencoders. It is to be noted

Table 2 LSTM autoencoders recall metrics

Historical maintenance records		Recall metrics %		
Vehicle machine state	Dates	Initial	Intermediate	Bad
Initial	x x x x x x x x x x x x x	84.98	85.35	85.35
Intermediate	x x x x x x x x x x x x	85.98	85.96	85.96
Bad	x x x x x x x x x x x x x	84.35	84.31	80.37
Initial	x x x x x x x x x x x x x	89.89	90.35	90.89
Intermediate	x x x x x x x x x x x x x	79.79	80.89	80.60
Bad	x x x x x x x x x x x x x	70.77	70.90	70.66
Initial	x x x x x x x x x x x x x	80.69	80.37	80.98
Intermediate	x x x x x x x x x x x x x	77.66	79.70	79.66
Bad	x x x x x x x x x x x x x	80.37	80.37	80.40

Table 3 LSTM autoencoders precision metrics

Historical maintenance records		Precision metrics		
Vehicle machine state	Dates	Initial	Intermediate	Bad
Initial	x x x x x x x x x x x x x	85.99	85.79	85.95
Intermediate	x x x x x x x x x x x x x	85.07	85.37	85.70
Bad	x x x x x x x x x x x x x	85.37	84.37	80.45
Initial	x x x x x x x x x x x x x	95.07	95.07	90.12
Intermediate	x x x x x x x x x x x x x	78.70	80.45	79.07
Bad	x x x x x x x x x x x x x	70.90	70.89	70.60
Initial	x x x x x x x x x x x x x	80.35	80.96	80.75
Intermediate	x x x x x x x x x x x x x	75.96	79.75	79.80
Bad	x x x x x x x x x x x x x	79.37	80.40	80.75

Table 4 LSTM autoencoders specificity metrics

Historical maintenance records		Specificity metrics		
Vehicle machine state	Dates	Initial	Intermediate	Bad
Initial	x x x x x x x x x x x x	95.35	95.98	95.35
Intermediate	x x x x x x x x x x x x	95.37	95.40	95.45
Bad	x x x x x x x x x x x x	95.70	95.95	95.96
Initial	x x x x x x x x x x x x	90.75	90.68	90.64
Intermediate	x x x x x x x x x x x x	90.35	90.84	90.98
Bad	x x x x x x x x x x x x	90.55	90.66	90.45
Initial	x x x x x x x x x x x x	90.79	90.89	90.95
Intermediate	x x x x x x x x x x x x	80.99	80.35	80.80
Bad	x x x x x x x x x x x x	80.85	80.99	80.90

Table 5 LSTM autoencoders F1 score metrics

Historical maintenance records		F1 score metrics		
Vehicle machine state	Dates	Initial	Intermediate	Bad
Initial	x x x x x x x x x x x x	90.41	90.70	90.66
Intermediate	x x x x x x x x x x x x	90.05	90.85	90.37
Bad	x x x x x x x x x x x x	85.87	80.37	80.36
Initial	x x x x x x x x x x x x	90.95	90.37	90.95
Intermediate	x x x x x x x x x x x x	80.55	80.35	80.80
Bad	x x x x x x x x x x x x	70.55	70.79	70.90
Initial	x x x x x x x x x x x x	80.90	80.80	80.50
Intermediate	x x x x x x x x x x x x	80.37	80.85	80.45
Bad	x x x x x x x x x x x x	80.36	80.37	80.35

that all dates represented in Table 8 are artificially created due certain restrictions imposed by business.

This difference characterizes and provides label for data type and status of vehicle's machine health accordingly. This observation is considered acceptable as company conducted maintenance activities on vehicle's machine before it is completely damaged. Thus, health status of vehicle is not bad during maintenance. As a result of this remaining useful life is extended.

Assuming all factors remain constant, vehicle's machine fatigue is proportional to distance it runs, time interval between two maintenance events is determined by vehicle's machine working time. Here, fatigue rate is considered constant and as shown in Table 8 vehicle's machine could become functional approximately in 3 to 5 days. Considering actual maintenance plan for period of approximately 6 months, a comparison has been performed between actual maintenance activity and maintenance suggested by proposed approach as shown in Table 8. As suggested by

Table 6 Accuracy results achieved (LSTM autoencoder vs. SVM variants)

Algorithm	States	Epochs 100-Batch10%	Epochs 70-Batch10%	Epochs 50-Batch10%
LSTM autoeucoder	Initial state	98.99	98.55	99.96
	Intermediate state	99.09	99.96	99.75
	Bad state	99.89	98.37	96.75
HMFSVM	Initial state	94.89	94.45	95.79
	Intermediate state	95.07	95.79	95.70
	Bad state	95.79	94.35	93.70
MFSVM	Initial state	90.79	90.65	90.79
	Intermediate state	90.12	90.89	90.89
	Bad state	90.89	90.45	90.99
SVM	Initial state	84.75	84.70	84.75
	Intermediate state	84.37	84.99	84.79
	Bad state	84.79	84.75	84.89

Table 7 LSTM autoencoders accuracy rates

Historical maintenance records		Accuracy rates %		
Vehicle machine state	Dates	Initial	Intermediate	Bad
Initial	x x x x x x x x x x x x	95.37	90.70	90.70
Intermediate	x x x x x x x x x x x x	80.45	89.37	80.45
Bad	x x x x x x x x x x x x	85.75	90.75	95.35
Initial	x x x x x x x x x x x x	99.09	80.95	70.45
Intermediate	x x x x x x x x x x x x	80.31	95.80	85.35
Bad	x x x x x x x x x x x x	95.54	85.70	85.55
Initial	x x x x x x x x x x x x	99.37	95.90	90.07
Intermediate	x x x x x x x x x x x x	75.31	90.80	85.37
Bad	x x x x x x x x x x x x	95.70	80.37	75.89

Table 8 Results from experiments

Start date	End date	Reason	Suggested end date	Days gained
27-12-2020	19-01-2021	Break down	17-01-2021	–
21-02-2021	05-03-2021	Preventive maintenance	07-03-2021	3
05-03-2021	17-03-2021	Preventive maintenance	21-03-2021	5

results, LSTM Autoencoders predicted vehicle's machine break down one day prior the actual event. The network results reflect that vehicle's machine is still in healthy state during maintenance activities. In a period of one year, maintenance activities take place every 30 days and vehicle's machine gains on average approximately 120 more days of life and 40% reduction in preventive stoppages.

The system developed tries to address several important issues faced by organizations in predictive maintenance. The LSTM Autoencoders based solution towards assessing vehicle's machine health condition have provided us appreciable results with respect to real world telematics datasets. However, this system has certain limitations which forms scope of future work. Additional experiments are required in order to further validate proposed hypothesis. This needs to be performed with larger datasets spanning over larger periods. With results obtained, redundant and stoppages in vehicle's machine operation can be significantly reduced. This in turn decreases cost towards maintenance operations. This method uses multiple neural networks in order to identify vehicle's machine status and RUL at higher resolutions. This can be error prone as there is a chance of fault classifications. This method uses from maintenance records' requirement for labeling datasets. This problem is coupled with need of large amounts of proper maintenance data such as vehicle's machine breakdowns. Another significant future work revolves around evaluation of this approach with respect to similar existing systems. Additional parameters revolving around wear and tear needs to be considered. The algorithm's performance can also be improved by optimizing network's hyperparameters. Also, a maintenance scheduling application can be developed involving both LSTM Autoencoders and transformer architectures.

5 Conclusion

Predictive maintenance of vehicle fleets has been an important topic of industrial internet of things research in academia and industry in lines with Industry 4.0 standards. In past few decades, companies have sent engineers and technicians on fixed schedules in order to address routine diagnostic inspections as well as preventive maintenance on vehicles deployed. However, this process is cost and labor intensive as well as error prone. Predictive maintenance maximizes vehicle lifespan and ensures optimized productivity. By harnessing vehicle garage data several problems are anticipated before failure happens. With this objective in view, LSTM Autoencoder based solution presented in this work can readily be deployed by organizations. This work assesses vehicle's health condition through estimating its RUL value. The system implementation is performed in Python. With these results redundant and preventive stoppages in vehicles are greatly reduced with decrease in maintenance operation costs. This has led towards deployment of advanced analytics and organizations gain ability to make decisions in real time. Companies can then perform necessary predictive maintenance and operations management with workflows triggered by analytics output. With these real-time, actionable insights customers are in a position to make predictive maintenance a reality based on advanced analytics across their deployment field. The future work involves addressing all stated limitations.

References

1. McKinsey & Company (2014) Connected car, automotive value chain unbound., Advanced Industries, McKinsey & Company
2. Intel (2016) Connected and immersive vehicle systems go from development to production faster, Intel
3. Intel (2016) Designing next-generation telematics solutions, White Paper In-Vehicle Telematics, Intel
4. Abel R (2017) Uber, Intel and IoT firms join coalition to secure connected cars
5. Vogt A (2017) Industrie 4.0/IoT vendor benchmark 2017: an analysis by Experton Group AG, An ISG Business, Munich, Germany
6. Predictive maintenance. https://en.wikipedia.org/wiki/Predictive_maintenance
7. Chowdhury M, Apon A, Dey K (eds) (2017) Data analytics for intelligent transportation systems. Elsevier
8. Chaudhuri A, Ghosh SK (2021) Predictive maintenance of vehicle fleets using hierarchical modified fuzzy support vector machine for industrial IoT datasets. In: Hybrid artificial intelligence systems, Lecture Notes in Computer Science, Springer, pp 331–342
9. Chaudhuri A (2021) Some investigations in predictive maintenance for industrial IoT solutions, Technical Report, Samsung R & D Institute Delhi India
10. Bampoula X, Siaterlis G, Nikolakis N, Alexopoulos KA (2021) Deep learning model for predictive maintenance in cyber-physical production systems using LSTM autoencoders. Sensors 21:972
11. Liu Z, Mei W, Zeng X, Yang C, Zhou X (2017) Remaining useful life estimation of insulated gate biploar transistors (IGBTS) based on a novel volterra K-nearest neighbor optimally pruned extreme learning machine (VKOPP) model using degradation data. Sensors 17:2524
12. Zhang C, Yao X, Zhang J, Jin H (2016) Tool condition monitoring and remaining useful life prognostic based on a wireless sensor in dry milling operations. Sensors 16:795
13. Aivaliotis P, Georgoulias K, Chryssolouris G (2019) The use of digital twin for predictive maintenance in manufacturing. Int J Comput Integr Manuf 32:1067–1080
14. Stavropoulos P, Papacharalampopoulos A, Vasiliadis E, Chryssolouris G (2016) Tool wear predictability estimation in milling based on multi-sensorial data. Int J Adv Manuf Technol 82:509–521
15. Oo MCM, Thein T (2019) An efficient predictive analytics system for high dimensional big data. J King Saud Univ—Comput Inf Sci
16. Bzdok D, Altman N, Krzywinski M (2018) Points of significance: statistics versus machine learning. Nat Methods 15:233–234
17. Huang CG, Huang HZ, Li YF (2019) A bidirectional LSTM prognostics method under multiple operational conditions. IEEE Trans Industr Electron 66:8792–8802
18. Liu C, Yao R, Zhang L, Liao Y (2019) Attention based echo state network: a novel approach for fault prognosis. In: Proceedings of ACM international conference on machine learning and computing, pp 489–493
19. Carvalho TP, Soares FA, Vita R, Francisco RDP, Basto JP, Alcalá SG (2019) A systematic literature review of machine learning methods applied to predictive maintenance. Comput Ind Eng 137:106024
20. Zonta T, Da Costa CA, Da Rosa Righi R, De Lima MJ, Da Trindade ES, Li GP (2020) Predictive maintenance in the industry 4.0: A systematic literature review. Comput Ind Eng 150:106889
21. Crespo Márquez A, De La Fuente Carmona A, Antomarioni S (2019) A process to implement an artificial neural network and association rules techniques to improve asset performance and energy efficiency. Energies 12:3454
22. Carbery CM, Woods R, Marshall AH (2018) A Bayesian network-based learning system for modelling faults in large-scale manufacturing, In Proceedings of IEEE international conference on industrial technology, pp 1357–1362
23. Wu N, Green B, Ben X, O'Banion S (2020) Deep transformer models for time series forecasting: the influenza prevalence case. arXiv arXiv:2001.08317

24. Vaswani A, Shazeer N, Parmar N, Uszkoreit J, Jones L, Gomez AN, Kaiser Ł, Polosukhin I (2017) Attention is all you need, In Proceedings of advances in neural information processing systems, pp 5999–6009
25. Guo Y, Zhou Y, Zhang Z (2020) Fault diagnosis of multi-channel data by the CNN with the multilinear principal component analysis. Measurement 171:108513
26. Chen CY, Huang JJ (2019) Double deep autoencoder for heterogeneous distributed clustering. Information 10:144
27. Murray B, Perera LP (2020) A dual linear autoencoder approach for vessel trajectory prediction using historical AIS data. Ocean Eng 209:107478
28. Rout AK, Dash PK, Dash R, Bisoi R (2017) Forecasting financial time series using a low complexity recurrent neural network and evolutionary learning approach. J King Saud Univ—Comput Inf Sci 29:536–552
29. Zhang J, Wang P, Yan R, Gao RX (2018) Deep learning for improved system remaining life prediction. Procedia CIRP 72:1033–1038
30. Malhi A, Yan R, Gao RX (2011) Prognosis of defect propagation based on recurrent neural networks. IEEE Trans Instrum Meas 60:703–711
31. Rezaeianjouybari B, Shang Y (2020) Deep learning for prognostics and health management: state of the art, challenges and opportunities. Measurement 163:107929
32. Gao S, Huang Y, Zhang S, Han J, Wang G, Zhang M, Lin Q (2020) Short-term runoff prediction with GRU and LSTM networks without requiring time step optimization during sample generation. J Hydrol 589:125188
33. Goodfellow I, Bengio Y, Courville A (2016) Deep learning. MIT Press, Cambridge, MA, USA
34. Wang Y, Zhao Y, Addepalli S (2020) Remaining useful life prediction using deep learning approaches: a review. Procedia Manuf 49:81–88
35. Yan H, Qin Y, Xiang S, Wang Y, Chen H (2020) Long-term gear life prediction based on ordered neurons LSTM neural networks. Measurement 165:108205
36. Fang Z, Wang Y, Peng L, Hong H (2021) Predicting flood susceptibility using long short-term memory (LSTM) neural network model. J Hydrol 594:125734
37. Shahid F, Zameer A, Muneeb M (2020) Predictions for COVID-19 with deep learning models of LSTM, GRU and Bi-LSTM. Chaos Solitons Fractals 140:110212
38. Bhuvaneswari A, Jones Thomas JT, Kesavan P (2019) Embedded bi-directional GRU and LSTM learning models to predict disasters on Twitter data. Procedia Comput Sci 165:511–516
39. Sayah M, Guebli D, Al Masry Z, Zerhouni N (2021) Robustness testing framework for RUL prediction Deep LSTM networks. ISA Trans 113:28–38
40. Sagheer A, Kotb M (2019) Unsupervised pre-training of a deep LSTM-based stacked autoencoder for multivariate time series forecasting problems. Sci Rep 9:1–16
41. Guo L, Li N, Jia F, Lei Y, Lin J (2017) A recurrent neural network-based health indicator for remaining useful life prediction of bearings. Neurocomputing 240:98–109
42. Wei W, Wu H, Ma H (2019) An autoencoder and LSTM-based traffic flow prediction method. Sensors 19:2946
43. Bao W, Yue J, Rao Y (2017) A deep learning framework for financial time series using stacked autoencoders and long-short term memory. PLoS ONE 12:e0180944
44. Gensler A, Henze J, Sick B, Raabe N (2016) Deep learning for solar power forecasting—an approach using autoencoder and LSTM neural networks. In: Proceedings of IEEE international conference on systems, man, and cybernetics (SMC), pp 2858–2865
45. Das L, Sivaram A, Venkatasubramanian V (2020) Hidden representations in deep neural networks: part 2. Regression problems. Comput Chem Eng 139:106895
46. Amirabadi M, Kahaei M, Nezamalhosseini S (2020) Novel suboptimal approaches for hyperparameter tuning of deep neural network. Phys Commun 41:101057
47. Preventive maintenance. https://en.wikipedia.org/wiki/Preventive_maintenance
48. Lathrop A (2017) Preventing failures with predictive maintenance: high-performance solutions using the Microsoft data platform, BlueGranite
49. Asghari V, Leung YF, Hsu SC (2020) Deep neural network-based framework for complex correlations in engineering metrics. Adv Eng Inform 44:101058

50. Yoo YJ (2019) Hyperparameter optimization of deep neural network using univariate dynamic encoding algorithm for searches. Knowl Based Syst 178:74–83
51. Rajapaksha N, Rajatheva N, Latva-Aho M (2019) Low complexity autoencoder based end-to-end learning of coded communications systems. arXiv arXiv:1911.08009
52. Tran KP, Nguyen HD, Thomassey S (2019) Anomaly detection using LSTM networks and its applications in supply chain management. IFAC-PapersOnLine 52(13):2408–2412
53. Nguyen HD, Tran KP, Thomassey S, Hamad M (2021) Forecasting and anomaly detection approaches using LSTM and LSTM autoencoder techniques with the applications in supply chain management. Int J Inf Manage 57:102282

A Comparative Study on the User Experience on Using Secure Messaging Tools

Blerton Abazi and **Renata Gegaj**

Abstract Considering the cyber-threats landscape over the past few years, the security community has repeatedly advocated the use of encryption for ensuring the privacy and integrity of the user's data and communications. As a result, we have witnessed a growth in number and usage of secure messaging tools. End-to-end encryption has been adopted by several existing and popular messaging apps. Email service providers have made efforts to integrate encryption as a core feature rather than treating it as optional. Recently, the pandemic situation has only reinforced this belief, causing us to rethink communication on a global scale, therefore put special emphasis on the security aspect of it. Despite the advances in research on usable security, the majority of these software still violates best practices in UI design and uses ineffective design strategies to communicate and let users control the security status of a message. Prior research indicates that many users have difficulty using encryption tools correctly and confidently. They either lack an understanding of encryption or the knowledge of its features. The paper investigates usable security design guidelines and models that lead to a better user experience for message encryption software. The study includes qualitative and quantitative research methods, conducted on a group of users, to discover the different factors that affect the user experience of encryption in everyday life. In the meantime, it also uncovers issues that current users are facing. Several recommendations and practical guidelines are outlined, these can be used by practitioners to make encryption more usable and thus increase the overall user experience of the software.

Keywords Information security and privacy · security · Email security · Encryption

B. Abazi (✉) · R. Gegaj
University for Business and Technology UBT, Prishtina, Kosovo
e-mail: blerton.abazi@ubt-uni.net

R. Gegaj
e-mail: rg33233@ubt-uni.net

© Springer Nature Switzerland AG 2022
R. Jiang et al. (eds.), *Big Data Privacy and Security in Smart Cities*,
Advanced Sciences and Technologies for Security Applications,
https://doi.org/10.1007/978-3-031-04424-3_7

1   Introduction

Before understanding the value of user-centered design, design decisions were mainly based on what designers' thought was suitable and what the client wanted to see. Interactions were based on what designers assumed to work best. The attention was heavily focused on aesthetics and the branding, with little to no thought of how the people who would use the product would feel about it. There was no research behind these decisions.

This decade has witnessed a transformation of digital design. The internet not only has it become widely used but websites and other applications have become so complex and feature-rich that, to be effective, they must have great user experience designs.

Additionally, users have been accessing websites in a number of ways: mobile devices, tablets, smart watches, browsers, different types of Internet connections etc.

With all of these rapid changes, the digital products that have consistently stood out were the ones that were pleasant to use. The driving factor of how professionals build digital products today has become the experience they want to give the people who will use these products. Alongside user experience, designers have also become aware of the importance of accessibility, for those who have special requirements, such as for screen readers and non-traditional input devices, also for users who have limited internet bandwidth or who have old mobile devices etc. Now design is based on understanding of users, tasks, and environments. It's always driven and refined by user-centered evaluation and addresses all aspects of user experience. User experience design combines market research, product development, strategy and design to create seamless user experiences for products, services and processes. It builds a bridge to the customer, making sure to better understand and fulfil users' needs and expectations. It takes into account every element that shapes the experience, how it makes the user feel, and how easy it is for the user to accomplish their desired tasks.

2   Literature Review

The exact definition of user experience, as outlined by the [1], is a "person's perceptions and responses resulting from the use and or anticipated use of a product, system or service."

In simple terms, user experience includes all the users' emotions, preferences, beliefs, perceptions, physical and psychological responses, behaviors and accomplishments that occur before, during, and after use. The ISO also lists three factors that influence user experience: the system, the user, and the context of use [2].

User experience (UX) focuses on having a deep understanding of users, what they need, what they value, their abilities, and also their limitations. It also takes into account the business goals and objectives of the group managing the project, brand, pricing, reports in media, etc.

In addition to the ISO standard, there exist several other definitions for user experience. Single experiences influence the overall user experience, the experience of a key click affects the experience of typing a text message, the experience of typing a message affects the experience of text messaging, and the experience of text messaging affects the overall user experience with the phone.

In short, user experience is how a person feels when interfacing with a system. The system could be a website/webapp, a mobile app or desktop software. Those who work on UX (called UX designers) study and evaluate how users feel about a system, looking at such things as ease of use, perception of the value of the system, utility, efficiency in performing tasks and so forth.

Factors that influence User Experience Use.

The core purpose of UX is to ensure that users find value in what a product is providing to them. There are many factors that determine the experience, none of them can stand on their own. This concept is best represented by the user experience honeycomb [3].

As it is presented in Fig. 1 to offer a meaningful and valuable user experience, the product must be:

Useful: The content should be original and fulfill a need
Usable: The product must be easy to use
Desirable: Image, identity, brand, and other design elements are used to evoke emotion and appreciation
Findable: Content needs to be navigable and locatable onsite and offsite
Accessible: Content needs to be accessible to people with disabilities
Credible: Users must trust and believe the product.

On this paper we will touch all these components but with a special focus on usability, as that is the aspect that most secure communication tools lack in.

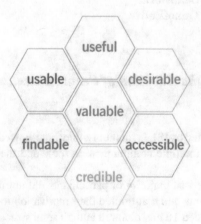

Fig. 1 The user experience honeycomb [4]

2.1 User Experience Research

User research focuses on understanding user behaviors, needs, and motivations through observation techniques, task analysis, and other feedback methodologies [5]. In short, it's the process of understanding the impact of design on an audience.

User research aims to find the barriers users face as they use the products, services, or processes by utilizing observational research methods to guide the design and development of a product.

User research applies to the entire product design process from ideation to the market release launch. It is an iterative process in which observation identifies problems and proposes solutions. Solutions are then prototyped and tested with the target user group.

The types of user research you can or should perform will depend on the type of site, system or app you are developing, your timeline, and your environment. User researchers work together with engineers, designers, product managers, and programmers in all stages of product development.

The User Research Goal and User Experience Research Method.

User research puts a project into context. By analyzing the data about users, designers and researchers can identify the problems users encounter during an interaction and turn them into actionable insights. By putting the user first and evaluating every design decision from their perspective, designers are able to create a more user-friendly experience that can lead users to return to a site, service, or product. There are numerous ways to conduct user research, while it's not practical to use all the methods on a single project, the best approach is to combine multiple research methods and from insights. The key question is what method to choose for a particular case.

The following chart helps us better understand 20 popular methods on a 3-dimensional framework with the following axes [6] (Fig. 2):

- **Attitudinal** versus **Behavioral**
- **Qualitative** versus **Quantitative**
- **Context of Use**.

2.2 Obstacles to the Adoption of Secure Communication Tools

This study was conducted by [8]. In this study, they interviewed 60 users and observed their experience with multiple communication tools and their perceptions of their security.

They found that the vast majority of participants did not understand the concept of end-to-end encryption, and that limited their motivation to use secure tools.

This research included 10 unstructured and 50 semi structured interviews.

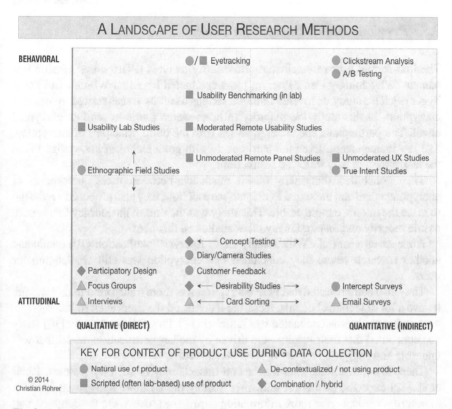

Fig. 2 A landscape of user research methods [7]

Although the participants had encountered usability issues with the communication tools, they found that usability was not the main obstacle to adopting secure tools.

The low motivation to use secure communication tools is due to several factors like small user bases and lack of understanding how secure communications work.

From their observation, users' goal to communicate with others overrides security.

According to the research, growing a user base for a new tool is difficult. Therefore, the best way to approach this problem is for security researchers to work with existing popular communication tools that have already been adopted by the majority of users instead of trying to improve the usability of new secure tools.

The technical security community must make efforts to develop a deeper understanding of what is important to users. Security should be presented in terms that users can understand.

2.3   *Previous Email Encryption Studies*

The first formal user research study of security features (PGP) on an email client named "Why Johnny Can't Encrypt" was conducted by Alma Whitten and Doug Tygar [9]. This study uncovered multiple serious usability issues related to message encryption. In this study the majority of users weren't able to send an encrypted email. The participants encountered issues with the basic principles of encryption, like key management. Even the participants with good technical knowledge failed the task although they followed instructions.

They concluded that there was a mismatch between users' perception of encrypting email and the actual PGP interface user flow, they also proposed a redesign to make the interface more usable. This study was the start of the journey for shaping usable security and inspired many other studies on this field.

After seven years of "Why Johnny Can't Encrypt" publication [10] conducted another research where they concluded that encryption was still challenging for users.

They also demonstrated that PGP encryption was more transparent than it should, it gave a lot of technical details, therefore it confused the users even more.

Another study was conducted by Garfinkel [11], they followed up on [10] study and concluded that their results were driven by the key certification model that was unusable and suggested replacing it for better results.

There is another study that touches on transparency of security features. Ruoti et al. [12] expected positive results after making the UI more transparent in terms of security by displaying more information during the process. On the contrary, that resulted in difficulties for users and in some cases, they even exposed their security.

The study concludes that a big barrier to users' understanding of email encryption are the "technical terms". The UI components, dialog boxes, menus, buttons and the certificates describe the process on terms that are not friendly for novice users or average users and the localization of this language adds another layer of difficulty.

Previous Studies on Instant Messaging Encryption Studies.

In a study conducted by Unger et al. [13] they compared a few platforms of secure instant messaging. They found that there are plenty of secure messaging services available on the market but they all struggle with making encryption approachable to users. The main blocker they found was insufficient user groups. Although all these platforms were trying to achieve anonymity but found out that can't be achieved without a large user base.

Another research was conducted by Assal et al. [14] who evaluated the usability of instant messaging services like Chat Secure and Tor and found usability bugs on both platforms.

The same research explored how user-friendly security indicators are. The findings show that users usually ignore or don't understand these security indicators but there are some common approaches to UI design that significantly improve the user experience. For example, the simple language without technical jargon, visible alerts, using colors animation/motion and the most important, security enabled by default.

Another research has been conducted by Evaluating E2E-Encryption in Popular IM Applications by Amir Herzberg and Hemi Leibowitz [10]. This study evaluates the implementation of end-to-end encryption in Briar messaging applications. It includes an expert review based on usable-security principles, and in quantitative and qualitative usability testing. The study shows that users of instant messaging applications are concerned about their privacy however they do not have the expertise to fully comprehend end-to-end encryption, they trust the application to provide this functionality and to clearly indicate the security status and any limitations, but the platforms resulted to be impractical to use and leave the users with an illusion of security.

An important topic named Expert and Non-Expert Attitudes towards (Secure) Instant Messaging was researched by Herzberg et al. [15].

This paper presents an online survey with 1510 participants and an interview study with 31 participants on secure mobile instant messaging. Their goal was to measure how much role security and privacy plays in people's decisions to choose a chat application. They found that security and privacy played an important role in the decision-making process only for some people, they were only seldom the primary factor, while how many people use the message platform, was the most important factor.

They also found that experts who had advanced knowledge about possible privacy and security risks, understood the tools on the same level as non-experts.

Some participants mainly non-experts perceived email as a secure communication tool because from their experience, the important information was sent through it.

3 Problem Statement

A good security feature should be invisible when you don't need it and helpful when you need it [16].

As we see from the studies mentioned above, this is hard to achieve. There are a lot of records that show how usability and user experience have affected the adoption of secure communication tools.

In general, user experience and security are perceived as opposites. There is a false belief that security focused tools, in this case communication tools, are targeted towards more "tech-savvy" users but in reality, every user needs protection, security and privacy. Every user should be able to use and understand these tools and their importance.

In practice, for example, setting up encryption keys, error messages etc. are complicated to understand for "average" users. Both the language and the design patterns seem to be dedicated only for IT security experts.

Secure communication tools need to be user friendly in order to be effective and used as they are intended.

Communication tools that have a backend that offers solid security but are complicated to use and hard to understand, will end up used incorrectly, therefore providing little or no protection.

We need more research to deeply understand how people make decisions about their privacy and security online, especially with communication tools. We need to observe how they interact with privacy and security aspects of the platform/app. Afterwards, we need to make an effort and come up with practical solutions on how to design interfaces that result in improved privacy and security.

To identify all these pain points and user needs, designers, user researchers, security engineers and every stakeholder has to be involved in ideating, designing, building and reviewing the interfaces, collaborating and keeping users in mind at all times.

Creating user centered design for secure communication tools it's not similar to designing for general communication tools, this case has similar design principles but with an added layer of complexity on top. There are many cases where designs principles are not set for this field, so designers have to come up with solutions as they go. This leads to chaos and uncertainty of the design.

A good example for this is the PGP encryption. The problem with it is not the interface components but the concept of encryption. Encryption in general is a more complex procedure which can be hard to understand for users in general. The concept is quite abstract and includes a language that is not used in users' everyday lives, for example public key, private key, key import/export, key ring etc. In addition to this, although there are many secure communication platforms there are no set "rules" and patterns for this specific design need. Therefore, all the platforms feel totally different even though they are built on the same principles. This leads to users being confused and makes it even more difficult for them to grasp the concept once and then expect the same procedures on every platform.

There are many experts that have been working on making these complex functionalities to users who don't understand the details of the system. This is a task-action based model that doesn't show the background of the working system. For example, "click the security button encrypt this message".

Users will always alter the intended use of a design, especially if they are under pressure. To make an effort for security they must first understand their data is under threat and it's their responsibility to protect themselves.

4   Research Methodology

4.1   Heuristic Review for Thunderbird Email Client

A heuristic evaluation is an inspection method that allows us to identify usability problems in a software design. Evaluators examine the interface and judge its compliance with recognized usability principles [17]. The main reason to conduct a heuristic review is to improve the usability and efficiency of a digital product. The user experience of an interface is greatly improved when these two components are delivered at a high quality.

This analysis can be performed at any stage of the design process, but it's not recommended for the very early stages. With new products/features, a heuristic evaluation is performed after wire framing and prototyping but before the UI development begins. Performing this evaluation too late will result in difficulty to make changes because of the time and cost increase.

Mozilla Thunderbird is a free and open-source cross-platform email client, personal information manager, news client, RSS and chat client developed by the Mozilla Foundation [18].

Thunderbird is offering a new feature that allows users to end-to-end encryption via OpenPGP. In the past this feature was achieved with the Enigmail extension [19], however, it is now being implemented as a functionality into core Thunderbird.

End-to-end encryption for email will ensure that only the sender and the recipients of a message can read the contents. Without this protection user's messages are more vulnerable and easier to monitor.

While this feature was still on beta version, it seemed the perfect opportunity to conduct a heuristic evaluation, to ensure it's usable before officially releasing it.

This study offers a consolidated report which not only identifies usability issues, but also recommendations on how to fix them when possible.

This review is based on Jakob Nielsen's 10 general principles for interaction design [20].

Areas Covered in this Review:

- Account Settings/End-To-End Encryption
- Key management
- Encryption settings
- Email
- Security/encryption for email
- Testing Environment
- Thunderbird Beta (78.0) for MacOS Catalina on a MacBook Pro 2018 and
- Ubuntu (18.04.4) on Lenovo ThinkPad T440s
- User Testing for Briar Messaging Application.

We conducted semi structured user interviews/usability testing sessions, to identify stumbling points in the current design implementation of Briar. Briar is a decentralized encrypted messaging service that is cross-platform. We chose Briar for this study due to its rapidly increasing popularity because of the security and privacy features it offers.

We ran the experiments with ten participants in total, representing a diverse range of users, with a special focus on at-risk users, that is, populations who for different reasons require enhanced security and privacy in their communication, for example journalists who work in risky regions. Their attitudes towards communicating via instant messaging is likely to be quite different from the ones of the general population. For this experiment, participants with a background in IT security or related technical fields were excluded, the reason being that there are already many studies that focus on these pools of participants, they also are more likely to not encounter major difficulties on grasping basic technical concepts or performing these tasks. All sessions were conducted remotely over video calls, where the participant shared their screen while being observed as they accomplished their tasks.

There were no time restrictions for task completion, but the goal was to keep the whole session within an hour, including the introduction and follow-up questions.

Each session was divided into two parts, the user interview and the usability testing.

In the first part, we evaluated the prior knowledge and beliefs that participants have in regard to privacy and communication in general. They were asked about their knowledge on security and privacy in messaging services, and how they feel about it but also touched on end-to-end encryption specifically. The questions were slightly different for each user depending on their background and their technological proficiency.

The second part of the session consisted of a traditional 1:1 usability testing.

On this part participants were given a few tasks to accomplish which were structured with the intention to explore the interfaces that are directly related to secure messaging, like settings, authentication, etc. Then they were asked to send an end-to-end encrypted message on Briar. We encouraged the participants to think aloud while completing the tasks and declare task completion when they think the message they sent was end-to-end encrypted. The participants were given a ten-minute break during the interview, to reduce interviewee fatigue and inattention. Each participant completed the tasks on a mobile phone with Briar already installed and prepared for testing.

The test started with me explaining briefly what usability testing is and why it's important.

4.1.1 User Testing for Transport Toggles

In this study were conducted five remote user testing sessions for the UI of transport toggles. The approach to testing this feature was interviewing combined with usability testing. This approach gives more context to why the user is having issues using the feature but also how this feature ties in with the overall user experience of Briar. Each session took 60–90 min. Participants were asked to think out loud while exploring the interface. During the first part of the session, the participant downloaded/installed the test apk and created an account, followed by questions: When they started using Briar, how they use it, frequency of usage, questions/tasks about different ways of connecting.

Participants of different backgrounds based on time and frequency of usage were chosen for this study (Fig. 3).

First Installed	Usage	Device	Location
Over a year ago	Daily	Pixel 3a	USA
6-12 months	2-3 times per year	Galaxy J2 Prime	Rio de Janeiro
Over a year ago	Weekly	LGv30	Austin Texas

Fig. 3 Participant Pool

4.2 System Settings Versus App Settings

In all cases, users had difficulties understanding how the switches in the navy drawer relate to the system-wide switches for enabling/disabling mobile data, Wi-Fi, and Bluetooth. They all had their assumptions on how that might work, based on common sense or they've read the documentation/looked online for help. Some tried to prove their assumptions are right by exploring different options. They tried switching the toggles on/off to see how they affect the system settings and then did the same for the system settings (turned on/off the Bluetooth, Wi-Fi) to see how they affect the app.

Even after exploring they weren't able to give a clear definition of how that works.

They weren't able to understand when they are connected to the internet and when they are offline. In one case the participant turned off the Bluetooth and Wi-Fi of the system and noticed that the "Tor" toggle remained enabled even when he tried to switch it off.

He wasn't sure if it was a bug or that's how it should work.

There is no clear indication of how Briar is currently connected. "I'm not sure if it's using Wi-Fi or mobile data. Does it switch automatically?".

Two participants had questions on how Tor works within the app. They weren't sure if they needed to have Tor installed on their phone as a prerequisite to be able to use it inside Briar. They explored the settings to find more information but couldn't find any.

5 Recommendations and Future Directions

What would be helpful in this case is improving the first-time experience using Briar. Prior to installing an application, most users don't gather information on how it works and technology behind it. They usually install it and figure it out as they go. Most of them don't take the time to ask for help even afterward (only one out of the three participants had read the docs and asked for help). In general, we want to make Briar easy to use and straight forward so that users will have everything they need to know within the app.

Because Briar is a unique application, that means that users can't compare it to other applications. That's why it's crucial to inform every user on how it works when they first start using the application. All of this would be possible if Briar would include an onboarding experience. The onboarding experience would give clear instructions using simple words and animations to explain What briar does How it works (How it connects to the internet, how it interferes with the system).

This would address most problems I described above. In addition to the onboarding experience, it's important to include a user guide so that users can access it when they need to get more information.

6 Conclusions

Although a long time has passed since the last formal study of encryption, the results of this study show that there are still many problems regarding the user experience.

There are several ways that this process could be improved. First, having a solid user onboarding would help first time users with step-by-step instructions. Second, a "non-technical" language of cryptography throughout the platform would help novice users grasp the concepts quickly. Third, a unified language to describe encryption, not only on one platform/software but in general. Finally, a design system that is made out of encryption specific components, will help users get familiar and understand encryption but also make it easier for designers to create and implement encryption on a platform.

The secure instant messaging apps on the other hand present very similar issues, a barrier for novice users. The main factor that kept users away from secure instant messaging apps was usability and the lack of features. Both these are expected to be similar to the most used apps like WhatsApp, Viber, Messenger etc. The expert and non-expert users' attitude towards privacy and security was different. Non-experts consider email as a secure communication tool because information they consider important is usually sent through it and it's what is mostly used in their work environments. For instant messaging apps, the company behind played the most important role. If the company is a well-known name, then they consider that application to be trustworthy.

A very important group on this research were at-risk users. These are users who for specific reasons require enhanced security and privacy in their communication. For example, journalists, whistleblowers, people living under government censorship. Their attitudes to security and privacy drastically changes compared to other users. In their case, security and privacy are crucial parts while they decide to install/use a messaging tool.

References

1. Ortiz Nicolás J, Aurisicchio M (2011) A scenario of user experience. In: ICED 11—18th international conference on engineering design—impacting society through engineering design, pp 182–193
2. Abro A, Sulaiman S, Mahmood AK, Khan M (2015) Understanding factors influencing user experience of interactive systems: a literature review. ARPN J Eng Appl Sci 10:18175–18185
3. Rosenbaum S, Glenton C, Cracknell J (2008) User experiences of evidence-based online resources for health professionals: user testing of The Cochrane Library. BMC Med Inform Decis Mak 8:34. https://doi.org/10.1186/1472-6947-8-34
4. Raemy J (2017) The international image interoperability framework (IIIF): raising awareness of the user benefits for scholarly editions
5. Al Qudah DA, Al-Shboul B, Al-Zoubi A, Al-Sayyed R, Cristea AI (2020) Investigating users' experience on social media ads: perceptions of young users. Heliyon 6:e04378. https://doi.org/10.1016/J.HELIYON.2020.E04378
6. Haenlein M, Kaplan A (2010) An empirical analysis of attitudinal and behavioral reactions toward the abandonment of unprofitable customer relationships. J Relatsh Mark 9:200–228. https://doi.org/10.1080/15332667.2010.522474
7. Rohrer C (2008) When to use which user experience research methods (Alertbox)
8. Abu-Salma R (2018) Exploring user mental models of end-to-end encrypted communication tools
9. Whitten A, Tygar JD (1999) Why Johnny can't encrypt: a usability evaluation of PGP 5.0. In: Proceedings of the 8th conference on USENIX security symposium, vol 8. USENIX Association, USA, p 14
10. Ruoti S, Andersen J, Zappala D, Seamons K (2015) Why Johnny still, still can't encrypt: evaluating the usability of a modern PGP client
11. Garfinkel SL (2008) Providing cryptographic security and evidentiary chain-of-custody with the advanced forensic format, library, and tools
12. Ruoti S, Andersen J, Monson T, Zappala D, Seamons K (2018) A comparative usability study of key management in secure email. In: Proceedings of the fourteenth USENIX conference on usable privacy and security. USENIX Association, USA, pp 375–394
13. Unger RA, Candelore BL, Pedlow LM, Jr (2002) Decoding and decryption of partially encrypted information
14. Assal H, Hurtado S, Imran A, Chiasson S (2015) What's the deal with privacy apps? A comprehensive exploration of user perception and usability
15. Herzberg A, Leibowitz H (2016) Can Johnny finally encrypt? Evaluating E2E-encryption in popular in applications. ACM Int Conf Proc Ser Part F 130652:17–28. https://doi.org/10.1145/3046055.3046059
16. Papernot N, About usable security. https://medium.com/@NicolasPapernot/about-usable-security-bb006cfbfda3
17. Penha M, Correia W, Campos F, Barros M (2014) Heuristic evaluation of usability—a case study with the learning management systems (LMS) of IFPE. Int J Humanit Soc Sci 4:295
18. MZLA Technologies Corporation: Mozilla Thunderbird. https://www.thunderbird.net/en-US/
19. Enigmail Project: Enigmail
20. www.nngroup.com/: 10 usability heuristics for user interface design. https://www.nngroup.com/articles/ten-usability-heuristics/

A Survey on AI-Enabled Pandemic Prediction and Prevention: What We Can Learn from COVID

Yijie Zhu, Richard Jiang, and Qiang Ni

Abstract COVID-19 pandemic spread quickly in Wuhan, China in December 2019. This destructive infection spread quickly all over the planet, causing enormous misfortunes of individuals and property. All over the planet, researchers, clinicians and legislatures are continually looking for new technologies to against the COVID-19 pandemic. The use of artificial intelligence (AI) innovation gives a better approach to battle the pandemic. This paper sums up the exploration and utilization of AI in forecast and avoidance of COVID-19 pandemics, and the possibility of AI innovation used to fight against the pandemic in the situation of smart city.

Keywords COVID-19 · Pandemic prediction · Pandemic prevention · Artificial intelligence

1 Introduction

A pandemic is a large-scale outbreak of a disease. It can increase mortality and incidence rate rapidly and cause major social, economic and political disturbances in a large and wide range. Due to the rising urbanization rate and more concentrated population distribution, the probability of pandemic has increased. So far, coronavirus has rapidly spread to almost all countries in the world. Since the primary instance of COVID-19 case in China was analyzed in Wuhan in December 2019, the pandemic circumstance has kept on spreading to the entire world, and the general wellbeing crisis of global concern was reported on January 30, 2020 [1]. Internationally, as of 11:16 am CEST, 6 July 2021, there have been 183,700,343 affirmed instances of COVID-19, including 3,981,756 deaths, according to WHO [2]. Coronavirus is an enormous class of infections, which can make sicknesses from normal cold intense or genuine illnesses. SARS-CoV-2 is an infection that causes COVID-19 pandemic, and is related with SARS-CoV-1, which caused severe acute respiratory

Y. Zhu (✉) · R. Jiang · Q. Ni
LIRA Center, Lancaster University, Lancaster LA1 4YW, UK
e-mail: y.zhu43@lancaster.ac.uk

© Springer Nature Switzerland AG 2022
R. Jiang et al. (eds.), *Big Data Privacy and Security in Smart Cities*,
Advanced Sciences and Technologies for Security Applications,
https://doi.org/10.1007/978-3-031-04424-3_8

syndrome (SARS). The common symptoms of coronavirus infection include persistent high fever, persistent dry cough, dyspnea, shortness of breath and other respiratory syndrome [3]. Since the COVID-19 pandemic have developed into a worldwide pandemic, many measures have been carried out to battle the spreading pandemic. Researchers, clinicians and legislatures are searching for new innovations to screen contaminated patients at various stages, control the spread of the infection, foster antibodies to treat tainted patients, and track the contact of tainted patients. Research shows that machine learning (ML) and artificial intelligence (AI) are promising advances, since they have better adaptability, quicker handling rate, reliability, and even better than individuals in specific medical care errands [4]. As of now, analysts have done much exploration on AI in assisting with anticipating and shield against COVID-19, and have made a ton of accomplishments and applications. This paper surveys the examination of AI in prediction and prevention of COVID-19 pandemic, including prediction and forecasting, screening and treatment, contact tracking, drugs and vaccination and some other research.

2 Prediction and Forecasting

COVID-19 and other pandemic diseases are spreading rapidly all over the world, posing a serious threat to global public health. Therefore, early and accurate prediction of pandemic will help to formulate appropriate policies and reduce losses. The prediction model can be utilized as a kind of perspective for planning new arrangements and assessing situation. It is accounted for that the COVID-19 pandemic is easy to spread [5]. Because of the vulnerability and intricacy of COVID-19 pandemic and its inconsistency in various nations, the performance of traditional epidemiological models has been challenged. Because of the critical distinction between the episode of COVID-19 and the new plagues (like swine fever, H1N1, Ebola and Zika), a few progressed epidemiological models have arisen to give higher exactness [6].

Many epidemiological models have been used to predict COVID-19 pandemics, such as susceptible-infected-removed (SIR), susceptible-exposed-infected-removed (SEIR) and susceptible-exposed-infected-removed–deceased (SEIRD). Similar to other epidemics, the general strategy behind SIR-based prediction model of COVID-19 outbreak is formed around the assumption that infectious diseases are transmitted through social contact. The SIR-based model assumes that the infection spreads from several groups of people, for example, susceptible population, exposed population, infected population, recovered population, dead population and immune population [7]. Literature [8] developed an altered SIR model of the COVID-19 disease in Italy. SEIR is similar to SIR, but the definition of incubation period is added, so it is more suitable for infectious diseases with a certain incubation period (Fig. 1). In literature [9], China's Wuhan population migration data in January 23rd and the latest COVID-19 epidemiological information were incorporated into the SEIR model. At the same time, artificial intelligence method was used to train the model according to the 2003 SARS information to predict the pandemic situation. SEIRD model adds dead

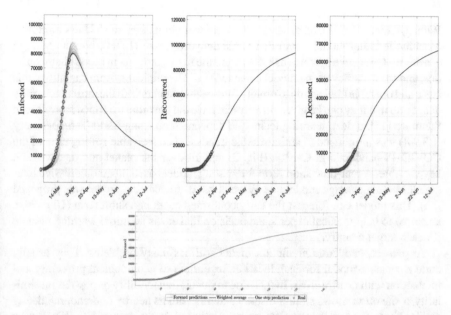

Fig. 1 Prediction of future values of infected, recovered, and deceased individuals in Italy using data up to March 30th, 2020 [8]

population. In literature [10], a SEIRD model is developed to examine and anticipate the spread of COVID-19 in Italy, which can better control the forecast caused by the blockade policy. In general, the SIR based model will be precise if the condition of social connection is steady. Second, class R should be determined accurately. To better estimate class R, different information sources can be coordinated with Sir based models, for example, CCTV, online media, portable applications and call records. However, using this model will still bring a lot of complexity and uncertainty. Since the development of SIR-based model involves a high degree of uncertainty, the generalization ability needs to be improved. The SIR-based model is only determined by a few parameters, so it is difficult to predict the complex infection pattern in a short time.

The accuracy of the above model is high in the long-term prediction, but it is weak in the short-term prediction, and is greatly affected by a single parameter. In order to overcome these shortcomings, machine learning method is applied to pandemic prediction. In literature [11], several machine learning methods are assessed in the undertaking of time series anticipating with one, three, and six-days ahead the COVID-19 combined affirmed cases in ten Brazilian states with a high daily infection. Among them, SVR has the highest accuracy. This study reinforces the short-term forecasting process, reminding medical experts and governments to respond to pandemics. A study [12] proposed another model utilizing a classifier XGBoost on clinical and mammographic factor datasets. To help the planning and strategies arranging of the medical system, the blood test data of 485 patients in

Wuhan, China was utilized to select the significant biomarkers of sickness mortality. Machine learning model found that lactic dehydrogenase (LDH), lymph and high sensitivity C-reactive protein (hs-CRP) were the main elements in blood characteristics, and the accuracy of the model came to 90%. The clinical system can utilize this basic and operable choice rule to rapidly foresee the patients with the most noteworthy danger, so that they can be given priority and have the potential to reduce mortality. In literature [13], a deep learning technique dependent on Long-Short-Term-Memory (LSTM) was introduced to anticipate the trends and conceivable halting time of the COVID-19 pandemic in Canada (Fig. 2) and all over the planet and discovered a key variable. A real-time short term forecasting model based on autoregressive integrated moving average model and Wavelet-based forecasting model was proposed [14]. This model can forecast the daily confirmed cases in a short term (10 days in advance) to help medical experts and decision makers as an initial warning module for each target country.

At present, pandemic prediction model still has many problems. They usually build a separate model for each location, regardless of geographical proximity and interaction with nearby areas. In fact, due to population mobility or population similarity, a site often shows similar disease patterns with its nearby or demographically similar sites. The existing models are mainly based on the data of COVID-19 case reports, but the missing report of these data or other data quality problems will affect the results of the model. Epidemiological models such as Sir use a set of differential equations to fit the whole curve of disease count. These models are only determined by a few parameters and are difficult to be used for short-term prediction [15]. On the contrary, the model based on deep learning can only make accurate prediction in a short time. Although some technologies combine these two methods [12], the existing models are difficult to provide accurate models in two times ranges.

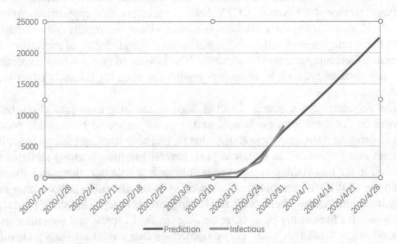

Fig. 2 Prediction of the LSTM model on current exposed and infectious cases [13]

3 Screening and Treatment

No matter what kind of disease, early detection is a significant undertaking for early treatment, especially for pandemics such as COVID-19. The fast determination and screening process assists with forestalling the spread of COVID-19 and other pandemic diseases, accelerate the speed of related diagnosis, and prevent the depletion of medical resources in a short time. The development of health care expert system is helpful to identify, screen and manage COVID-19 carriers, and improve the efficiency of pandemic prevention.

In the process of diagnosis, AI technology is often used in computed tomography (CT), X-ray and clinical blood sample data to enhance the diagnosis and screening process of identified patients. In general, medical staff can use radiation images such as CT scan or X-ray for diagnosis and screening. However, during the period of high incidence of COVID-19 pandemic, the efficiency of related equipment is difficult to cope with the growing number of suspected and confirmed patients. In literature [16], a quick and powerful analysis strategy for COVID-19 dependent on AI technology is proposed. Ten notable convolutional neural networks were utilized to recognize the patients with and without COVID-19 disease. In all networks, ResNet-101 and Xception show the best result. The accuracy of ResNet-101 was 99.51% and Xception was 99.02%. In the research of [17], they use Darknet model as the classifier of You Just Look Once (YOLO). Through the examination of chest X-beam pictures, the model can give exact determination to two classes (COVID-19 and no finding) and multi-classes (COVID-19 and no finding and pneumonia).

In addition to the above methods, blood diagnosis is also a good method to detect COVID-19. In the study [18], they found 11 key indicators that can be used to identify COVID-19. In this study, random forest algorithm was used to extract these 11 related indicators, the overall accuracy was 95.95%, and the specificity was 96.97%. These indicators can be used as the identification tools of COVID-19 to help medical experts diagnose quickly.

The main purpose of AI application in screening and treatment is to carry out rapid diagnosis. Timely and early detection can lessen the spread of COVID-19, and allow for clinical specialists to do the next diagnosis, so as to save more lives and reduce medical costs. However, most studies use a single classification algorithm for single or multiple data.

4 Contact Tracing

Assuming an individual is affirmed to be infected with COVID-19, the following step is to prevent the wider spread of the disease through contact tracing. COVID-19 is mainly spread among people through saliva, droplets or contact, with strong transmission ability. Based on the characteristics of COVID-19, contact tracing is extremely valuable in obstructing the spread of the infection. The task of contact tracing is to recognize people who have recently contacted with patients infected with COVID-19 in order to avoid further spread. Under normal circumstances, using this technology can break the current COVID-19 transmission chain, thus limiting the spread scale of pandemic and gradually controlling the pandemic. Different countries use different technologies, such as Bluetooth, global positioning system (GPS), mobile tracking data, system physical address and so on, to design mobile applications with contact tracing. These applications are mainly used to collect personal data, such as action tracks and public transport ride records. Based on these data, AI tools will be able to track people who have a certain degree of contact with users of the app and are likely to be infected with the virus. As shown in Table 1, article [19] lists the countries with such ML and AI based contact tracing application capabilities. However, there are still limitations in the field of privacy and data security related applications. These applications will collect a lot of personal information, including action trajectory and public transport ticket information. There is a risk of personal information leakage in these processes. However, as the pandemic continues to spread around the world, the COVID-19 pandemic has become a global health emergency, and almost every country has launched its own contact tracking applications. Considering the necessity of pandemic prevention and control, it is necessary to formulate some unified standards.

5   Drugs and Vaccination

Since the outbreak of COVID-19, the development of vaccines has brought the pandemic prevention and control to a new stage. Vaccination has effectively controlled the spread of the pandemic in some countries, but with the constant variation of COVID-19, especially the emergence and spread of Delta virus, the early developed vaccine has reduced its immune function to some extent. Therefore, a method for developing drugs and vaccines against COVID-19 is important, and AI technology can provide an idea for this.

In literature [20], a new model, namely molecular transformer drug target interaction (MT-DTI), was proposed to solve the demand for antiviral drugs that can treat COVID-19 virus. In this research, they utilized a pre-trained medication targeting model based on deep learning to recognize commercially accessible medications that can work on SARS-CoV-2 infection protein. The model adopts the deep learning algorithm of 3C-like proteinase and 3410 drugs approved by FDA. The results showed that the antiretroviral drugs for the treatment and prevention of Human Immunodeficiency Virus were the best compounds. Research [21] noted the similarities between COVID-19 and Ebola virus and Zika virus in terms of difficult treatment. They proposed a method combining computational screening with docking application and machine learning to select auxiliary drugs to study SARS-CoV-2.

Research [22] proposed a method, that is, using artificial intelligence technology, through two databases, to set up an AI platform to distinguish the old drugs that are effective against the coronavirus. The results showed that 8 kinds of drugs could

Table 1 Contact tracing application used by countries

Location	Name	Tech
Australia	COVIDSafe	Bluetooth
Austria	Stopp Corona	Bluetooth, Google/Apple
Bahrain	BeAware	Bluetooth, Location
Bangladesh	Corona Tracer BD	Bluetooth, GPS
Canada	COVID Alert	Bluetooth, Google/Apple
China	Chinese health code system	Location, Data mining
Czech Republic	eRouska	Bluetooth
Denmark	Smittelstop	Bluetooth, Google/Apple
Estonia	HOIA	Bluetooth, DP-3 T, Google/Apple
Finland	Koronavilkku	Bluetooth, Google/Apple
France	TousAntiCovid	Bluetooth
Germany	Corona-Warn-App	Bluetooth, Google/Apple
Hungary	VirusRadar	Bluetooth
India	Aarogya Setu	Bluetooth, Location
Indonesia	PeduliLindungi	Bluetooth, Location
Italy	Immuni	Bluetooth, Google/Apple
Japan	COCOA	Bluetooth, Google/Apple
New Zealand	NZ COVID Tracer	Bluetooth, QR codes
Norway	Smittestopp	Bluetooth, Google/Apple
Poland	ProteGO Safe	Bluetooth
Saudi Arabia	Tabaud	Bluetooth, Google/Apple
South Africa	COVID Alert SA	Bluetooth, Google/Apple
Switzerland	SwissCovid	Bluetooth, DP-3 T, Google/Apple
UK	NHS COVID-19 App	Bluetooth, Google/Apple
Vietnam	BlueZone	Bluetooth

inhibit the proliferation of feline infectious peritonitis (FIP) virus in Fcwf-4 cells. In addition, five other drugs were also found to be active in the practice of AI methods. According to the previous experience of patients, these old drugs can be easily used to fight against the SARS-CoV-2 pandemic if they are proved to be effective against SARS-CoV-2.

AI technology can more effectively predict the reusability of COVID-19 drugs for existing old drugs, and greatly reduce the risk in the process of developing more cost-effective drugs. In the case of COVID-19 rampant emergencies, the application of AI can enhance the drug development process by reducing the time interval between the discovery of complementary therapies and drugs.

6 Other AI Application of COVID-19

According to the World Health Organization (WHO), wearing masks in public places is one of the effective protection methods. COVID-19 is most likely to spread through respiratory droplets, so it is an effective way to spread virus transmission by ensuring that people wear masks generally. Study [23] proposes a face detection model based on the combination of deep learning and some machine learning methods. The model consists of two components. The first component is used for feature extraction with Resnet50. The second component uses decision tree, support vector machine (SVM) and ensemble algorithm to classify human faces. Three datasets are used in the study. The results show that the method is able to accurately detect the mask. In study [24], they introduce a new facemask-wearing detection model by combining image super-resolution and classification deep neural networks. The results show that the model can recognize the wearing status of face mask with high accuracy, and the classification accuracy of SRCNet under kappa is 98.70% (Fig. 3).

Due to the rapid spread of the pandemic and the limited medical resources, many countries have the medical system overload operation, unable to detect, admit and treat all patients in time. Overburdened health care systems and poor disease surveillance systems may not be able to cope with the outbreak of COVID-19, which requires tailored strategic responses to these environments. Literature [25] recommends a low-cost self-detection and tracking system coupled with blockchain and artificial intelligence for COVID-19 and other similar diseases. Rapid deployment and appropriate implementation of the system have the potential to curb the spread of COVID-19 and associated deaths, especially in environments where laboratory infrastructure is lacking. With this system, people can collect samples by themselves and use the point-of-care equipment for rapid detection.

With the continuous spread of COVID-19 pandemic, many Italian people have carried out behaviors inconsistent with the protective health measures, such as not wearing masks, leaving the isolation place without permission, holding multi person parties, etc. These behaviors will lead to the further spread of COVID-19, which is not conducive to the prevention and control of pandemic. Research [26] uses machine learning to anticipate which people are bound to follow defensive measures.

Fig. 3 Mask identification examples in real situations [24]

The results showed that the score of behavioral compliance was lower than that of effective perception. The mediating effect of risk perception and citizen attitude on self-efficacy is not significant. These results are helpful to guide the countries affected by COVID-19 to carry out information / advertising activities more effectively.

Along with the coronavirus pandemic, in addition to the casualties caused by pandemic itself, mass fear and panic phenomena also constitute a crisis, incomplete and inaccurate information contributed to the formation of this phenomenon. Therefore, only by better understanding the information of COVID-19 and measuring public sentiment can decision-makers implement information transmission and decision-making. In study [27], they used coronavirus specific tweets and R statistical software, as well as emotion analysis software package to identify public emotions related to influenza pandemic. This study provides two machine learning classification methods and compares their effectiveness in tweets of different lengths in the context of text analysis. This study helps to understand the source of public fear and solve the problem of mass fear.

7 Pandemic Prevention in Smart City

In recent years, globalization, urbanization, population explosion and other social phenomena have profoundly changed people's way of life. About half of humankind lives in metropolitan regions and settlements. It can be expected that this proportion will rise further in the coming decades. For the sustainable development of cities, environmental and health challenges are serious problems, which may threaten the livability and quality of life of citizens. Especially with the outbreak of COVID-19 pandemic, urban areas have suffered more losses than rural areas due to over concentration of population, insufficient allocation of medical resources and poor crowd management. All of this requires that all stakeholders properly and effectively address multidisciplinary issues and integrate citizens into the planning of sustainable city. Therefore, a highly integrated system is essential.

Smart city can collect a large amount of user centered and technology driven data in real time through intelligent Internet devices, and optimize resource allocation from traffic, logistics, travel control and other aspects to cope with highly infectious pandemics such as COVID-19.

AI and big data can provide evidence-based prediction for local decision makers. Literature [28] investigates the outbreak of COVID-19 from the perspective of cities, and shows how smart cities and smart networks use high-quality and improved standardized protocols to share data in emergency situations, so as to better manage such situations.

The continuous COVID-19 pandemic has exposed the limitations of the existing smart city planning. Therefore, it is very important to develop a system and architecture that can provide a mechanism to limit the further spread of the pandemic. Literature [29] proposes a data-driven smart city sustainable development framework based on deep learning, which can timely respond to the COVID-19 pandemic through large-scale video surveillance. They use three real-time object detection models based on deep learning to detect people in the video captured by monocular camera (Fig. 4).

Study [30] observed the impact of Swedish blockade strategy by monitoring the change of urban noise level. The data used were recorded during one year-long noise level survey on the veneer of a building at a bustling metropolitan intersection in Stockholm, Sweden.

8 Conclusion and Discussion

In this study, we summarize the research and application of different artificial intelligence in the fields of COVID-19 prediction and forecasting, screening and treatment, contact tracking, drugs and vaccination. The results show that AI technology plays an important role in the prediction and prevention of COVID-19. The combination of technologies in various fields and AI greatly improves the reliability of prediction, screening, contact tracking and drug/vaccine development. Among them, deep learning technology has a good performance in detection and recognition tasks. Deep learning algorithm has greater potential, robustness and advanced in other algorithms. However, most of the models cannot show their performance in the real world, but they still provide a reference for fighting against COVID-19 pandemic. The data of coronavirus patients all over the world will be a good source for AI researchers

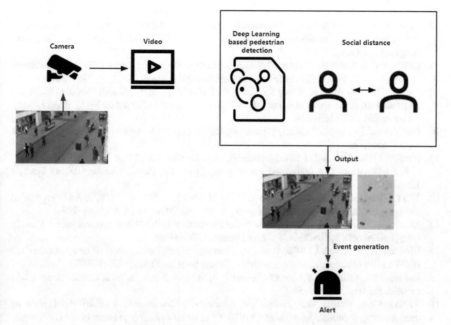

Fig. 4 Illustration of the real-time video object detection system

to develop automatic diagnostic tools, formulate treatment strategies for coronavirus patients, and deal with possible pandemics in the future. As more authentic dataset will be available in the future, the more accurate AI-based application can be implemented.

References

1. Sohrabi C, Alsafi Z, O'Neill N, Khan M, Kerwan A, Al-Jabir A, … Agha R (2020) World Health Organization declares global emergency: a review of the 2019 novel coronavirus (COVID-19). Int J Surg 76:71–76
2. WHO (World Health Organization) (2021) Coronavirus disease (COVID-2019) dashboard. https://covid19.who.int. Accessed 6 July 2021
3. Lai CC, Shih TP, Ko WC, Tang HJ, Hsueh PR (2020) Severe acute respiratory syndrome coronavirus 2 (SARS-CoV-2) and coronavirus disease-2019 (COVID-19): the epidemic and the challenges. Int J Antimicrob Agents 55(3):105924–105924
4. Davenport T, Kalakota R (2019) The potential for artificial intelligence in healthcare. Future Healthc J 6(2):94–98
5. Ivanov D (2020) Predicting the impacts of epidemic outbreaks on global supply chains: a simulation-based analysis on the coronavirus outbreak (COVID-19/SARS-CoV-2) case. Transp Res Part E 136(136):101922–101922
6. Koolhof IS, Gibney KB, Bettiol S, Charleston M, Wiethoelter A, Arnold A-L, … Shiga T (2020) The forecasting of dynamical Ross River virus outbreaks: Victoria, Australia. Epidemics 30:100377

7. Koike F, Morimoto N (2018) Supervised forecasting of the range expansion of novel non-indigenous organisms: alien pest organisms and the 2009 H1N1 flu pandemic. Glob Ecol Biogeogr 27(8):991–1000
8. Calafiore GC, Novara C, Possieri C (2020) A modified SIR model for the COVID-19 contagion in Italy. In 2020 59th IEEE conference on decision and control (CDC), pp 3889–3894
9. Yang Z, Zeng Z, Wang K, Wong SS, Liang W, Zanin M, … Mai Z (2020) Modified SEIR and AI prediction of the epidemics trend of COVID-19 in China under public health interventions. J Thorac Dis 12(3):165–174
10. Piccolomini EL, Zama F (2020) Monitoring Italian COVID-19 spread by an adaptive SEIRD model. MedRxiv
11. Ribeiro MHDM, da Silva RG, Mariani VC, dos Coelho, LS (2020) Short-term forecasting COVID-19 cumulative confirmed cases: perspectives for Brazil. Chaos, Solitons Fractals 135:109853
12. Yan L, Zhang HT, Goncalves J, Xiao Y, Wang M, Guo Y, … Zhang M (2020) An interpretable mortality prediction model for COVID-19 patients. Nat Mach Intell 2(5):283–288
13. Chimmula VKR, Zhang L (2020) Time series forecasting of COVID-19 transmission in Canada using LSTM networks. Chaos, Solitons Fractals 135:109864
14. Chakraborty T, Ghosh I (2020) Real-time forecasts and risk assessment of novel coronavirus (COVID-19) cases: a data-driven analysis. Chaos, Solitons Fractals 135:109850
15. Roberts MG, Andreasen V, Lloyd A, Pellis L (2015) Nine challenges for deterministic epidemic models. Epidemics 10:49–53
16. Ardakani AA, Kanafi AR, Acharya UR, Khadem N, Mohammadi A (2020) Application of deep learning technique to manage COVID-19 in routine clinical practice using CT images: results of 10 convolutional neural networks. Comput Biol Med 121:103795
17. Ozturk T, Talo M, Yildirim EA, Baloglu UB, Yildirim O, Acharya UR (2020) Automated detection of COVID-19 cases using deep neural networks with X-ray images. Comput Biol Med 121:103792
18. Wu J, Zhang P, Zhang L, Meng W, Li J, Tong C, … Zhu J (2020) Rapid and accurate identification of COVID-19 infection through machine learning based on clinical available blood test results. MedRxiv.
19. MIT. Covid tracing tracker—a flood of coronavirus apps are tracking us. Now it's time to keep track of them. https://www.technologyreview.com/2020/05/07/1000961/launching-mittr-covid-tracing-tracker/. Accessed 30 June 2021
20. Beck BR, Shin B, Choi Y, Park S, Kang K (2020) Predicting commercially available antiviral drugs that may act on the novel coronavirus (SARS-CoV-2) through a drug-target interaction deep learning model. Comput Struct Biotechnol J 18(18):784–790
21. Ekins S, Mottin M, Ramos PRPS, Sousa BKP, Neves BJ, Foil DH, … Southan C (2020) Déjà vu: Stimulating open drug discovery for SARS-CoV-2. Drug Discovery Today 25(5):928–941
22. Ke YY, Peng TT, Yeh TK, Huang WZ, Chang SE, Wu SH, … Song JS (2020) Artificial intelligence approach fighting COVID-19 with repurposing drugs. Biomed J 43(4):355–362
23. Loey M, Manogaran G, Taha MHN, Khalifa NEM (2021) A hybrid deep transfer learning model with machine learning methods for face mask detection in the era of the COVID-19 pandemic. Measurement 167(167):108288
24. Qin B, Li D (2020) Identifying facemask-wearing condition using image super-resolution with classification network to prevent COVID-19. Sensors 20(18):5236
25. Mashamba-Thompson TP, Crayton ED (2020) Blockchain and artificial intelligence technology for novel coronavirus disease-19 self-testing. Diagnostics (Basel, Switzerland) 10(4):198–198
26. Roma P, Monaro M, Muzi L, Colasanti M, Ricci E, Biondi S, Napoli C, Ferracuti S, Mazza C (2020) How to improve compliance with protective health measures during the COVID-19 outbreak: testing a moderated mediation model and machine learning algorithms. Int J Environ Res Public Health 17(19):7252
27. Samuel J, Ali GGMN, Rahman MM, Esawi E, Samuel Y (2020) COVID-19 public sentiment insights and machine learning for tweets classification. Inf Int Interdisc J 11(6):314

28. Allam Z, Jones DS (2020) On the coronavirus (COVID-19) outbreak and the smart city network: universal data sharing standards coupled with artificial intelligence (AI) to benefit urban health monitoring and management. Healthcare 8(1):46
29. Shorfuzzaman M, Hossain MS, Alhamid MF (2021) Towards the sustainable development of smart cities through mass video surveillance: a response to the COVID-19 pandemic. Sustain Cities Soc 64:102582
30. Rumpler R, Venkataraman S, Göransson P (2020) An observation of the impact of COVID-19 recommendation measures monitored through urban noise levels in central Stockholm, Sweden. Sustain Cities Soc 63:102469

Blockchain Based Health Information Exchange Ecosystem: Usecase on Travellers

Fatima Khalique, Sabeen Masood, Maria Safeer, and Shoab Ahmed Khan

Abstract Various international initiatives have been launched in 2021 to generate digital vaccination certificates to address challenges faced by travellers exposed during COVID19. While COVID19 has impacted multiple industries where alternate solutions have emerged at competing rates, airline industry in particular have been hit hard due to highly regulated protocol. Owing to its unique characteristics of transparency and decentralization, block chain technology provides an opportunity to create smart solutions for travel industry. We present a solution that describes a scaled, blockchain-based infrastructure for exchanging COVID-19 pandemic travellers' history and vaccination status in a secure manner. The proposed approach ensures the travellers are coronavirus-free and check their COVID 19 history and vaccination status across the borders and immigration in particular. The framework employs a permission blockchain and Proof of Authority to check vaccination status on airports using the smart contract. It provides a distributed infrastructure for national and international healthcare systems, to check digital vaccination certificate and individual vaccination history and their verification by relevant stakeholders, such as airport securities, health authorities, governments, border control authorities and airlines.

Keywords Vaccination certificate · COVID-19 · Blockchain · Proof of authority

F. Khalique (✉)
Department of Computer Science, Bahria University, Islamabad, Pakistan
e-mail: fkhalique.buic@bahria.edu.pk

S. Masood
Department of Computer and Software Engineering, National University of Sciences and Technology, Islamabad, Pakistan
e-mail: sabeen.masood@ceme.nust.edu.pk

M. Safeer
Department of Management Sciences, Sir Syed Case Institute of Technology, Islamabad, Pakistan

S. A. Khan
Centre for Advanced Research in Engineering, Islamabad, Pakistan

© Springer Nature Switzerland AG 2022
R. Jiang et al. (eds.), *Big Data Privacy and Security in Smart Cities*,
Advanced Sciences and Technologies for Security Applications,
https://doi.org/10.1007/978-3-031-04424-3_9

1 Introduction

COVID-19 was declared a pandemic on March 11th, 2020 by the World Health Organization. This virus is caused by the corona virus 2 (SARS-CoV-2) acute respiratory syndrome, which was initially discovered in the Chinese city of Wuhan at the end of 2019 [1, 2]. Since then, the disease has easily spread around the world. For its fast spread, almost every country in the world is at risk, and existing medical services are overwhelmed. The only guaranteed way to stop the virus from spreading is to maintain personal hygiene and practice social distance. Aside from the obvious health benefits, the financial impact is already evident in many nations throughout the world [3]. Indeed, drastic and occasionally contentious measures to prevent the spread, such as curfew and social distancing, have already changed our daily life. Many countries throughout the world have implemented rigorous quarantine for all infected residents, travel restrictions, and social distancing measures like as public closings to prevent mass gatherings. As one of the most socially interactive activity, tourism has been hit hard during the pandemic.

Many countries prohibited non-essential travel during the initial wave of the epidemic to slow the spread of Covid-19. As the COVID-19 vaccination campaign progresses, countries have started to lift international travel restrictions and allow travellers from specific destinations to cross the border. Different countries developed their own border screening protocols for travelling, which were often dependent on the traveller's origin country [4]. Some countries agree with the concept of a "risk-free certificate" or "immunity passport" that would allow individuals who have antibodies against Covid-19 to travel and be protected from re-infection. However, until today, there has been no proof that the antibodies generated by Covid-19 infection can prevent subsequent infection or for how long [5]. Immunity certifications are commonly referred to as "immunity passports," however they are more accurately described as immunity-based licenses. These policies present serious problems about justice, stigma, and counter productive incentives, but they also have the potential to advance individual liberty and enhance public health [6]. COVID-19 vaccination certificates or cards are also issued to those who have been vaccinated and contain information such as personal information, health facility, immunization dates, and dosage [7]. Smallpox, yellow fever (which is necessary for travellers in and out of endemic countries), and diphtheria-tetanus-pertussis for children have all been successfully managed with vaccine in the past [8]. Traditional immunization certificates or cards, on the other hand, are susceptible to forgery [9], corruption, modifications, and are difficult to read by non-health professionals, as well as being vulnerable to weather conditions such as rain [10].

The International Health Regulations have set international travel requirements for the Covid 19 vaccination [11]. Several airlines require a negative COVID-19 test before boarding [12], and travellers heading back to their location are needed to isolate within their own expense while waiting for the results of their COVID-19 test [13]. Another important tool is traveller's testing after they reach a destination location. The findings could lead to the screening of travellers who arrived by the

Fig. 1 Blockchain
electronic contract process

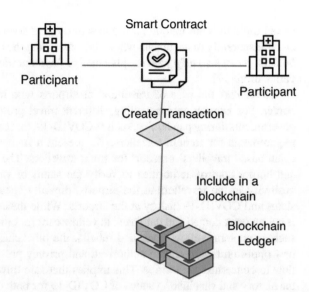

same mode of transportation, reducing the risk of COVID-19 transmission at the border. Currently, certificates such as COVID19 test results and vaccination proof are all paper-based, making them vulnerable to falsification and inefficient processing and verification in many countries [14]. There is currently an absence of a robust data surveillance network that might effectively supply healthcare entities with the information they require regarding COVID-19 patient history and vaccination certificates worldwide. So there is a critical need for a global digital tracking method to check vaccination status of individual that travelled overseas since the coronavirus epidemic. Since the coronavirus outbreak, governments have begun introducing technologies to check vaccination certificate before boarding the plane at international airports However, there is no internationally digitally recognized tracking method to check whether an infectious or tested person has travelled internationally since the corona virus epidemic. So, digital proof of vaccination is being needed to check the traveller Covid-19 history and vaccination status at airports.

Digital COVID-19 vaccination certificates must be remotely accessible, fortified, protect people's privacy and secure. Blockchain has emerged as a promising technology in a variety of fields, including technology, food, banking, energy sector, security, health, authentication in the smart grid, agriculture and automation industry [15], and has been dubbed the "next great invention" since the internet. It establishes business-to-business or peer-to-peer networks with end-to-end encryption using a decentralized, distributed ledger. A smart contract is a computer protocol used in blockchain to expedite, execute, or verify the digital execution or negotiation of a contract, and it allows transactions to take place without the intervention of third entities as shown in Fig.1. Encryption can take place on both sides privately or, more commonly, privately on one end and publicly on the other. The concept "blockchain" refers to a chain of digital "blocks," each containing a set of data or records and is linked digitally with unique hash codes through encryption [16]. It can efficiently and transparently manage data while also preserving the confidentiality of all stakeholders. It can also help with supply chain or payment transaction authorization and verification [17].

We address the case of travellers' at airports who travel inter state or across border. For international travelling, different travel protocols are implemented by governments during pandemics, such as COVID-19, that establish a series of entrance requirements for foreign travellers. We present a framework that based on block chain based travellers' passport for smart travellers. The passport will help airport and border control authorities to verify the status of vaccination and COVID-19 positivity in the travellers at the airports. digitally checks travellers' vaccination status and COVID-19 history at the airports. While the concept of health passports is not un precedented and date back to yellow card for other communicable diseases such as cholera, yellow fever and rubella, the blockchain technology has created new opportunities to ensure authorized and privacy preserving health information flow to concerned authorities. This implies that safe travelling as well as to assess the history and vaccination status of COVID-19 for both domestic and international travellers within the country and across borders using blockchain. In this paper, Sect. 2 covers the related work. Problem statement along with the existing methodology, followed by a proposed novel framework are outlined in Sect. 3. Section 4 discusses the implications of the proposed methodology. Section 5 concludes the paper.

2 Related Work

Blockchain technology has enormously gained popularity since its first application in crypto currency Bitcoin. It provides a diverse set of opportunities in a variety of fields. Researchers and scientists have been looking at its application in the health sector in the last few years [18]. Nevertheless, the first use of blockchain in the health sector was in 2011, when a unified database for doctors, pharmacists, nurses, and other healthcare stakeholders was built [19]. To combat the effects of COVID-19, many organizations and global companies are incorporating blockchain into their respective solution space. The primary goal of using blockchain is to bring together all of the trusted and verified data sources. One of the blockchain's distinct feature is its ability to constantly authenticate data in real-time, which is essential in the battle over COVID-19 [17]. Data privacy and security issues must also be addressed when health data exchange is discussed owing to its sensitive [20]. Zhang et al. [21] analyzed several healthcare blockchain application use cases. They have discussed the importance of a blockchain in healthcare and how it is helpful in healthcare design.

Smart contracts and blockchain technology can aid the healthcare industry by expediting the processes through efficient data management that reduces data loss and prevents data falsification on the ledger [22]. Several researches discuss application of blockchain from different perspectives in wake of COVID-19 as well as concerns, challenges and barriers associated with its use [23, 24].

Numerous immunity passport papers [7, 25, 26] employ self-sovereign identity (SSID) and verifiable credentials systems to provide people complete control over their personal information, which is a desirable feature. These solutions, however, imply establishing a method, such as a mobile app, for individuals to exercise this control, which may not be viable for everyone around the world. In [7], authors create a mobile application to make COVID-19 test and vaccination certificates easier to issue and validate. This approach only allows for the sharing of private data, including test and vaccine certificates, and therefore does not allow for the distribution of public information. It uses verifiable credentials to make electronic certificates that allow users to save confidential documents on their phones. One limitation proposed method is that it needs people to produce a QR code for their credentials when using their smartphones. This method only allows for minimal document checking without a smartphone, which is insufficient for confirming test and vaccination certifications. Furthermore, it is unclear how this method handles the possibility of fraudulent information using the mobile application. Ethereum smart contracts have also been used to create a blockchain system for issuing and publishing COVID-19 vaccine and test certificates [27]. These smart contracts carry out tasks like adding a test canters, updating patient data and generating a test result. The majority of the information on the blockchain is notifications regarding the execution of smart contracts. Private data is saved off-chain through proxy re-encryption mechanisms using the InterPlanetary File System (IPFS). Furthermore, it relies on SSID to provide users complete control over their personal data. A public blockchain is used that is more susceptible to attacks and has worse scalability and throughput [26]. In [28], the authors developed a blockchain-based tracking system for exchanging COVID-19 data from various sources. By using Ethereum smart contracts to track reported data via credible sources, the proposed approach reduces the spreading of manipulated or fraudulent data. This method is only ideal for sharing public statistics such as the number of new and recovered cases, but not for transferring private data such as tests and vaccination certificates.

In an attempt to address the travelling challenge during the COVID-19 epidemic, the authors presented an architecture to support digital health passports that employed a private blockchain in [29]. Three primary components make up the framework: (1) local healthcare facilities that issue digital health passports to travellers, (2) blockchain members with only read access to the blockchain (such as airport security, airline companies, and border control authorities), and (3) health service authorities (at the national level) with complete access to the blockchain, including authorized issued digital health passports, mainly to evaluate whether a person is the owner of a correct digital health passport. Furthermore, the framework makes use of mobile phones to protect an individual's privacy. The suggested architecture has privacy vulnerabilities, such as the disclosure of passenger test history, as well as the reality that any blockchain member might potentially get a person's information, which is irrelevant to the particular blockchain member.

Other approaches discuss utilizing blockchain to distribute individual vaccination records [30, 31]. The authentication process involves using date of birth, gender, and iris template. However, these approaches publicly publishes the individuals' health record information in the blockchain where records are available to anyone who obtains the identifying information, for example user's iris template, date of birth and gender. In [32], the authors formulate a secure antibody certificate system named SecureABC, which utilizes a standard public-key signature scheme to assure certificate authenticity and binding. This effort has the drawback of relying on an individual's photo and name for verification. In [33], the authors proposed a SPIN framework that uses a permissioned blockchain to share the information and is successfully battling against COVID-19 spread over the world. The SPIN architecture allows the sharing of public data via a permissioned ledger that is available to all peers, and private data via private data collections.

3 Proposed Transformation Centered Around Smart Contract

Covid-19 has had the unprecedented effect on air transport industry. The pandemic has affected aviation stakeholders including airports, airlines, businesses and passengers both economically and socially. Since first administered vaccine shot from 2020 and continued efforts of countries to increase vaccination rates of population, the airports have gradually moved from strict travel bans to partially restrictive regulation. However, in order to ensure safe travels and less disruptive border control restrictions, a coordinated approach is required that may allow resuming smooth travel patterns as desired by both travellers and air industry. This coordination involves all stakeholders that are previously unrelated to each other as shown in Fig. 2.

Fig. 2 Block chain allows entities to share information without having any direct business relationship to each other

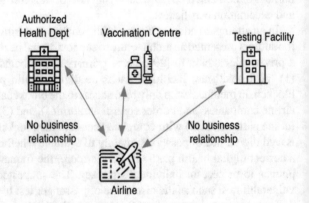

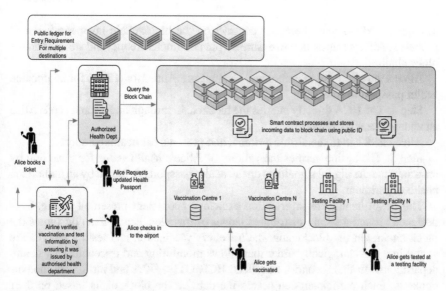

Fig. 3 Proposed health information exchange ecosystem for travellers

The proposed framework works through multiple features when implemented through block chains. The block chain technology allows the stakeholders to access true state of shared health data for a traveller. This data includes all the transactions related to parameters for travelling under restrictions such as traveller's vaccination status, type, date of administration, COVID test date, type and results. Blockchain offers different degrees of privacy and anonymity, transparency and immutability of the traveller's health records. The identity of individuals getting vaccinated or tested is replaced through a public key that is a hash key which cannot be decoded to identifiable information. The information provided by Health Care Facility (HCF) and Covid Testing Facility (CTF) gets stored on the blockchain in the form of transactions by triggering smart contracts. The block chain based health information exchange for ecosystem shown in Fig. 3 using a health passport for traveller's constitutes the following system when defined through classical Alice example.

Step 1: Alice gets vaccinated at HCF. The data is stored in the HCF existing clinical data base. Appropriate data fields and Alice's public key is sent to the blockchain layer through an API.

Step 2: The smart contract on the block chain layer receives and processes Alice's getting vaccinated transaction to ensure all pre conditions are met and stores it on the block chain. Alice later decides to travel and books a ticket. The airline books a ticket and based on destination information, retrieves travel requirements for Alice and make them known to her.

Step 3: Alice visits a CTF and gets tested for the required COVID test. The data is stored in the CTF existing data base. Appropriate data fields including information on test and its results along with Alice's public key is sent to the blockchain layer through an API. Again the smart contract on the block chain layer receives and processes Alice's transaction to ensure all pre conditions are met and stores it on the block chain.

Alice send a request to authorized Health Care Authority (HCA) for an updated health passport to travel.

Step 4: The HCA directly queries the blockchain through an API and sends Alice an updated health passport.

Alice checks in to the airline terminal and presents her health passport.

Step 5: The airline verifies the validity of Alice's health status for travelling by ensuring that the digital signatures on the health passport are issued by an authorized health department.

The travel healthcare information exchange ecosystem presented in Fig. 3, is designed around a secure distributed ledger of travellers health that constitutes the block chain. On the blockchain, the travellers vaccination and testing records are stored in a chronologically order that allows monitoring and traceability by health departments. In this scenario, travellers, HCF, CTF , HCA and airline are the participants. Each participant can access the data on the block chain based on their business requirements and controlled through their rights over the data. Specifically the use of block chain for sharing COVID19 vaccination and test related information gives the following advantages: All of the tests and vaccination history records are stored on the block chain, that can be accessed and monitored according the constantly changing COVID restrictions regarding, expiry duration, type of vaccine valid, booster administration etc.; In the event, that an individual decides to travel, or tested positive, the information can be accessed on the block chain via individual's digital health passport and medical history can be traced; Every transaction on the block chain is digitally signed by the creator HCF and CTF with collision resistant hash function. Each transaction is added to the blockchain through Merkle tree and hash functions. These techniques ensures the immutability of the transactions on the block chain that provides security against malicious attacks.

4 Discussion

We discuss the travellers use case under COVID-19 restrictions as the regulations in both health and travel industry are crucial in terms of public health interventions, privacy and economy. The proposed approach allows authentication and validation of the identities of both travellers and certificate issuing authorities while remaining privacy compliant. The proposed solution utilizes the unique characteristics of block chain to achieve consensus through proof of authority. Blockchain technology gives several advantages over traditional processes to health care providers, medical researchers, and individuals [17].

The proposed approach allows smart travels by making it easier to securely exchange information across various healthcare systems along with traveller approval. Blockchain improves the security of electronic travellers' health data and communication links, particularly when sharing and accessing Covid-19 data. Additionally, this approach allows securing digital vaccination certificates of COVID-19. Electronic vaccination certificates are verified and validated across the borders using blockchain.

Blockchain also gives a secure potential solution for health information exchange. The presented use case comes with inherent challenges of privacy and security of sensitive health data. In parallel, blockchain offers several advantages in its own design to address these challenges. Blockchain offers immutability of transactions that means that once data is added to the block it cannot be tampered with. In addition, even if a node is compromised in the blockchain, the rest of the flow works without interruption and the threat is not propagated to other nodes. The privacy of the individual data on block chain can be discussed through differentiating between data owner, creator and manager. In this use case, data creator are the HCF and CTF where traveller gets vaccinated and tested respectively. The authorized HCA manage, monitor and protect the data shared from multiple sources. The owner of this data remain the individual itself with right to share this information outside the blockchain. However, once the data is placed on the block chain, it cannot be modified by creator, owner or manager. This also allows preservation of data integrity with interoperability. In addition, the read write access of different participants can be defined on a. block chain where the individual owner may not be allowed to write on the block chain. In the context of privacy and security for health data, it is worth mentioning that while some requirements of sharing vaccination and tests records maybe guaranteed through inherent properties of blockchain architecture and workflow, there maybe some requirements that need to be addressed through other technologies. For example, confidentiality, authentication and fine grained access control to the data can be further enhanced through key management and encryption. The use case presented in this paper can serve as a foundation towards creating a scalable, decentralized and secure solution using block chain where partial privacy and security requirements maybe addressed through other available technologies. Health care block chains in general are an innovative solution towards sharing healthcare transactions in a secure healthcare network.

5 Conclusion

The use of blockchain in healthcare provides many potential use cases owing to unique characteristics of block chain. Health data exchange, access and control are closely regulated in the blockchain technology. The blockchain contains distributed ledger that keeps transactions on nodes throughout the network. Blockchain addresses the security and authenticity aspects in the health data exchange, as it cannot be tampered with or altered. In this paper, we proposed a novel blockchain-based solution to globally check and track travellers' COVID-19 history and their vaccination status at airports for seamless authenticated and safe travel experience. This proposed framework presents a solution for travellers using the smart contract.

Travel industry has faced serious economic crises due to different COVID-19 proto-cols implemented by the governments for travellers. The proposed approach aim to create a smart solution for health authorities, airport securities, governments, aviation industries, and residents to make essential decisions for travellers.

References

1. World Health Organization (WHO) Timeline of WHO's response to COVID-19, 2020, https://www.who.int/news/item/29-06-2020-covidtimeline
2. Rachmawati I, Shishido K (2020) Travelers' motivations to travel abroad during Covid 19 outbreak. Int J Appl Sci Tourism Events 4(1):1–11
3. OECD Economic Outlook, Interim report September 2020, OECD, 2020
4. Bastani H, Drakopoulos K, Gupta V, Vlachogiannis I, Hadjicristodoulou C, Lagiou P, Magiorkinis G, Paraskevis D, Tsiodras S (2021) Efficient and targeted COVID-19 border testing via reinforcement learning. Nature 599:108–113
5. World Health Organization (2020) "Immunity passports" in the context of COVID-19. Geneva
6. Persad G (2020) The ethics of COVID-19 immunity-based licenses ("Immunity Passports")
7. Eisenstadt M, Ramachandran M, Chowdhury N, Third A, Domingue J (2020) Covid-19 anti-body test/vaccination certification there's an app for that. IEEE Open J Eng Med Biol 1:148–155
8. Wagner AL (2019) The use and significance of vaccination cards. Hum Vaccines Immunother 15(12)
9. Marhold K, Fell J (2021) Electronic vaccination certificates: avoiding a repeat of the contact-tracing 'format wars. Nat Med 1–2
10. Mbunge E (2021) Emerging technologies and COVID-19 digital vaccination certificates and passports. Public Health Pract (Oxf) 2:100136
11. World Health Organization. Interim position paper: considerations regarding proof of COVID-19 vaccination for international travellers, https://www.who.int/news-room/articles-detail/interim-position-paper-considerations-regarding-proof-of-covid-19-vaccination-for-international-travellers
12. Kelleher RS (2021) Get ready for needing a negative Covid-19 test to fly within the U.S. Forbes https://www.forbes.com/sites/suzannerowankelleher/2021/02/09/get-ready-for-needing-a-negative-covid-19-test-to-fly-within-the-us/?sh=297595282eed (2021)
13. Government of Canada (2021) Flying to Canada: COVID-19 testing for travellers—travel restrictions in Canada—Travel.gc.ca
14. Lemmon A (2021) Re-shaping the future of identity through user-owned verifiable credentials. In: IEEE international conference on blockchain and cryptocurrency (ICBC)
15. Rimsan M (2020) COVID-19: a novel framework to globally track coronavirus infected patients using blockchain. Int Conf Comput Intell, ICCI 70–74
16. Vervoort D, Guetter CR, Peters AW (2021) Blockchain, health disparities and global health. BMJ Innov 7:506–514
17. Shah H, Shah M, Tanwar S, Kumar N (2021) Blockchain for COVID-19: a comprehensive review. Pers Ubiquit Comput
18. Zheng Z, Xie S, Dai H, Chen X, Wang H (2017) An overview of blockchain technology: architecture, consensus, and future trends. In: IEEE international congress on big data (BigData congress), pp 557–564
19. Gupta R, Tanwar S, Tyagi S, Kumar N (2019) Tactile-internetbased telesurgery system for healthcare 4.0: an architecture, research challenges, and future directions. IEEE Netw 33(6):22–29
20. Litchfield AT, Khan A (2019) A review of issues in healthcare information management systems and blockchain solutions; CONF-IRM. Available online: https://aisel.aisnet.org/confirm2019/1/

21. Zhang P, Schmidt DC, White J, Lenz G (2018) Blockchain technology use cases in healthcare. Advances in computers, vol 111. Elsevier, Amsterdam, The Netherlands, pp 1–41
22. Siyal A, Junejo A, Zawish M, Ahmed K, Khalil A, Soursou G (2019) Applications of blockchain technology in medicine and healthcare: challenges and future perspectives. Cryptography 3(3)
23. Kumar T, Ramani V, Ahmad I, Braeken A, Harjula E, Ylianttila M (2018) Blockchain utilization in healthcare: key requirements and challenges. In: Proceedings of the 2018 IEEE 20th international conference on e-Health networking, applications and services (Healthcom), Ostrava, Czech Republic, 17–20 Sept 2018
24. Kalla A, Hewa T, Mishra RA, Ylianttila M, Liyanage M (2020) The role of blockchain to fight against COVID-19. IEEE Eng Manag Rev 48:85–96
25. Hernández-Ramos JL, Karopoulos G, Geneiatakis D, Martin T, Kambourakis G, Fovino IN (2021) Sharing pandemic vaccination certificates through blockchain: case study and performance evaluation. arXiv arXiv:2101.04575
26. Wüst K, Gervais A (2018) Do you need a blockchain? In: Proceedings of the crypto valley conference on blockchain technology (CVCBT), Zug, Switzerland, 20–22 June 2018. IEEE: New York, pp 45–54
27. Hasan HR, Salah K, Jayaraman R, Arshad J, Yaqoob I, Omar M, Ellahham S (2020) Blockchain-based solution for COVID-19 digital medical passports and immunity certificates. IEEE Access 8:222093–222108
28. Marbouh D, Abbasi T, Maasmi F, Omar IA, Debe MS, Salah K, Jayaraman R, Ellahham S (2020) Blockchain for COVID-19: review, opportunities, and a trusted tracking system. Arab J Sci Eng 45:1–17
29. Angelopoulos CM, Damianou A, Katos V (2020) DHP framework: digital health passports using blockchain. arXiv, arXiv:2005.08922
30. Singh A, Raskar R (2020) Verifiable proof of health using public key cryptography. arXiv, arXiv:2012.02885
31. Chaudhari S, Clear M, Tewari H (2021) Framework for a DLT based COVID-19 passport. arXiv, arXiv:cs.CR/2008.01120
32. Hicks C, Butler D, Maple C, Crowcroft J (2020) SecureABC: secure AntiBody certificates for COVID-19. arXiv, arXiv:cs.CR/2005.11833
33. Alabdulkarim Y, Alameer A, Almukaynizi M, Almaslukh A (2021) SPIN: a blockchain-based framework for sharing COVID-19 pandemic information across nations. Appl Sci 11(18):8767. https://doi.org/10.3390/app11188767

Video-Based Heart Rate Detection: A Remote Healthcare Surveillance Tool for Smart Homecare

Thomas Harrison, Zhaonian Zhang, and Richard Jiang

Abstract A novel approach to extract a heart rate signal from video footage consisting of a five stage processing pipeline is presented. Two extraction methods were used to obtain a heart rate. The first used the Fast Fourier transform to estimate an average heart rate by peak frequency analysis in the frequency distribution and estimated heart rates with a MAE as small as 2.32 BPM. This MAE value is smaller than those found by previous research which used PPG signals and BCG signals to extract a heart rate. The second approach used the Short-time Fourier transform to produce a time series of heart rate estimation which, when compared to accepted ground truths produced a covariance value of up to 0.9206335. Using a hybrid CNN-LSTM model an ECG-like signal was extracted from time-series heart beat waveforms. The resultant ECG-like signal displayed some of the characteristic ECG traits however it was not stable across the entire time period. Potentially, such a non-invasive heart monitoring can serve as a remote healthcare surveillance tool for smart homecare.

Keywords Smart homecare · Artificial intelligence of things · Heartrate monitoring

1 Introduction

The ability to monitor and analyse a persons heart rate may give insight into their current health conditions; this will give medical professionals the ability to detect early stages of heart conditions such as Cardiovascular Disease (CVD). In specialist environments the heart rate can be measured using several different technologies such as an electrocardiogram (ECG) and blood measurement in a controlled environment [1]. There have been many attempts to detect and monitor heart rate signals without the use of medical grade equipment such as contact-based systems which utilise the same framework as oximeters (photoplethysmography) [2]. Alongside contact-based systems that utilise photoplethysmography (PPG) there have been many studies that

T. Harrison · Z. Zhang · R. Jiang (✉)
LIRA Center, Lancaster University, Lancaster, UK
e-mail: r.jiang2@lancaster.ac.uk

© Springer Nature Switzerland AG 2022
R. Jiang et al. (eds.), *Big Data Privacy and Security in Smart Cities*,
Advanced Sciences and Technologies for Security Applications,
https://doi.org/10.1007/978-3-031-04424-3_10

have estimated a persons heart-rate through remote PPG analysis [3]. Touch-based PPG analysis and remote PPG analysis have their benefits since they are cost-effective and provide user comfort however they have limitations. Contact-based systems are susceptible to motion which may negatively influence the clarity of the produced PPG signal. The ability to accurately monitor a subjects heart rate using cost-effective methods that do not require the use of a medical practitioner to operate will provide accessibility to the general public.

Alongside medical application the ability to monitor a persons heart-rate can extend to devices such as lie detectors [4] and to distinguish emotions [5] through heart-rate variability.

A series of three minute long videos were taken of five different, healthy subjects. Three different video resolutions were used: 1920×1080 px (25 FPS), 2705×1520 px (29.97 FPS) and 3840×2160 px (14.99 FPS). Alongside the recording factors two cases of subject movement were consider, a case where the subject remained still and a second case were the subject was asked to move naturally to better assess the limitations of each heart-rate extraction model.

Extraction of a heart-rate used a five stage pipeline: region of interest (ROI) selection and pixel tracking, noise removal, signal isolation and component synchronisation. This processing was performed prior to heart-rate estimation and ECG extraction.

The ROI was selected as the upper chest region and feature points were defined inside of this region using Shi-Tomsai minimum eigenvalue algorithm [6] and tracked using a Kande-Lucas (KLT) tracking algorithm. The noise removal step compares the frequency distribution obtained using the FFT on the pixel displacements in the ROI with the frequency distribution obtained from pixels on a stationary background object. The signal isolation alleviated frequencies that were considered as noise by implementing a Kaiser-Bessel finite impulse response (FIR) bandpass filter with a passband of $[1-1.67]$Hz (with adjustments for special cases). The wave synchronisation step ensures that all of the processed position-time heart-rate waveforms aligned in the time domain by comparing the phase difference to an 'ideal' base waveform.

After the processing pipeline the heart-rate was estimated using two fundamental approaches. The first approach used the Fast Fourier transform. Two different methods were used to estimate the heart-rate from the FFT. Peak frequency analysis in the FFT frequency distribution and an alternate temporal approach using the prominent number of peaks in the heart-beat waveform.

The second approach generated a time series waveform using the STFT windows and used peak frequency analysis within each window to generate a series of heart-rates. The window size of the STFT was optimised by using a mean absolute error (MAE) loss function to minimise the window size without impacting the accuracy of the predicted heart-rates. The estimated heart rates were compared to ground 'accepted' truths obtained using a health smart watch worn by each subject which recorded their hear-rates across the duration of the video.

The resultant pixel waveforms from the FFT and STFT analysis were used in an attempt to create an ECG-like signal. The first of two approaches used the sinusoidal waveforms generated from each pixel alongside a CNN-LSTM hybrid model.

2 Literature Review

Previous research using non-medical grade equipment is mostly contained within two categories:

- Non-contact heart rate detection through use of remote detection. These methods often use video footage of a subject.
- Contact methods that often use fingertip sensors.

There are two well researched methods of extracting a heart-rate through non-contact and contact approaches that were considered whilst researching this project:

- PPG/RIPPG
- BCG

Photoplethysmogramy (PPG) is a form of contact heart-rate detection by using a wearable device such as a fingertip sensor on a smart phone or wearing a heart-rate sensing watch. This method commonly uses an infra red light emitting diodes (LEDs) or alternatively a green light emitting diode to illuminate the skin and measure the contrast in the reflected and absorbed light caused by the volume-metric change in blood flow. This change in blood volume is caused by the constriction and dilation of the capillaries. RIPPG is related at its core principle but takes a different approach. RIPPG uses video footage to measure variance in the green, red and blue light reflections of the skin.

Ballistocardiography (BCG) measures the mechanical movement caused by the mechanical action of a heart beat. This movement can be detected by means of remote imaging which commonly involve measure vertical head movement from the influx of blood at each heart beat. There are also reports of BCGs used as a form of contact heart detection by using a water tube and a pressure sensor to measure the ballistic forces caused by a heart-beat. The following research focuses on remote BCG as it is most closely related to the method proposed in this paper.

The remained of the literature review is split into two sections. Section 2.1 covers previous research into PPG and RIPPG methods. Section 2.2 focuses on remote BCG anlysis.

2.1 Previous Research Which Use PPG/RIPPG

N.H.Mohd Sani et al. obtained the heart rate using both the Fast Fourier transform (FFT) and number of peaks in the time-domain signal. Their results found that the FFT analysis provided a slightly lower heartrate signal than using the peak of the time-domain signal, howevever they demonstrated no ground truth heart-rates for comparison [7]. A similar approach used by Ratna et al. utilised FFT Analysis on a Photoplethysmograph (PPG) signal to classify normal and abnormal heart rates [8].

Li et al. used a four stage pipeline to generate a PPG signal with facial video footage. This pipeline consists of ROI detection and feature point tracking. The first stage in the pipeline was defining the ROI which used a Viola-Jones face detector to identify the facial region. This was followed by a Discriminative Response Map Fitting (DRMF) [9] method to find the co-ordinates of facial landmarks. The region that contains eyes was was eliminated to exclude inference caused by blinks and the facial boundary was indented to exclude the possibility of non-facial pixels being tracked. The feature points were identified using the Shi-Tomsai algorithm [6] and tracked using the Kanade-Lucas-Tomasi (KLT) algorithm [10]. The second stage involved Illumination Rectification in an attempt to remove the interference caused by rigid head movement. This method incorporated Normalised Least Mean Square adaptive filtering methods to reduce the motion artifacts. The third stage involved attempting to resolve interference due to non-rigid head movement which may be caused by facial expressions. This step began by dividing the pulse signal into equal segments and calculating the standard deviation of each segment. The top 5% of segments which contained the largest standard deviation were considered as contaminated and removed. The final stage in their pipeline introduced a series of temporal filtering approaches. These filters operated in the frequency range of [0.7, 4]Hz and consisted of a detrending filter, a moving average filter and a Hamming based FIR filter. There proposed method was successful in comparison to previous work with an average error rate of 6.87% across 527 tested samples. Their limitation lay with large angle head rotation where the feature points were lost and unable to be tracked which caused accuracy loss [11].

Bush 2016 used remote PPG analysis on the facial region within video footage to measure the heart-rate. The proposed method tested the algorithms capabilities by subjection to two different movement states, where the subject is asked to move naturally and when the subject remains still. Both results were promising with an error of 3.4 ± 0.6 bpm whilst the subject was still and 2.0 ± 1.6 bpm when the subject was moving [12]. These errors were calculated by comparing their estimated heart-rate values compared to ground truth heart-rates from a contact PPG finger tip sensor. Whilst the heart-rate errors are low and provide accurate results the method itself still has its limitations since the method was only tested in well lit situations and the bounding box which encloses the subjects facial region was not tested to quantify how 'clutter' could impact the performance.

For non-contact heart-rate detection Blackford et al. studied the effects of video resolution and frame rate on the mean error distribution; they found no significant difference between 60FPS and 30FPS when these were reduced from 120FPS. Similar results were found with a reduced resolution 329×246 px when compared to the original resolution of 658×492 px [13].

Studies on remote heart-rate detection include the work of Wang et al. which used a three stage pipeline consisting of face video processing, signal extraction for the face blood volume pulse (BVP) and using the extracted signals to compute a heart rate [14].

Poh et al. developed a non-contact cardiac pulse measurement method using video footage with a webcam [15]. Using a tracking algorithm, pixels in the facial region were tracked alongside using Blind Source Separation (BBS) by Independent Component Analysis (ICA) to estimate the colour channel temporal signals. This process was used to recover the recovered blood volume pulse (BVP) signals and compared to an FDA-approved finger blood volume pulse (BVP). Poh et al. improved on this method in later studies by refining the 'Region Of Interest' (ROI) step. This advancement was reducing the ROI to 60% width and full height of the box. This reduction in size is to remove corner pixels that may not be located in the facial region. Additionally, a temporal bandpass filter (128-point Hamming window operating between 0.74 Hz) was included to refine and smooth the PPG signal [16].

Feng et al. estimated the heart rate of a subject by means of remote imaging photoplethysmography (RIPPG). RIPPG monitors the variation in luminance of the skins pixels caused by the cardiac pulse [17]. Feng et al. used a Viola-Jones face detection algorithm to select the ROI as the facial region accompanied with a Kanade–Lucas–Tomas tracking algorithm (KLT) to track feature points. A raw RIPPG signal is obtained by averaging the green and red channels in the pixels that are in the ROI. A heart-rate frequency is estimated on this GRD signal and use an Adaptive Bandpass Filter (ABF) to eliminate noise so that a more comprehensible signal is produced.

2.2 Previous Research Which Use BCG

Balakrishnan et al. extracted a subjects heart rate by measuring slight head movements caused by the influx of blood at each heart-beat [18]. Balakrishnan et al. used the Viola Jones face detector [19] from OpenCV [20] to identify the facial region and tracked the coordinates of the feature points using the OpenCV Lucas Kanade tracker. They opted for the middle 50% width-wise and 90% height-wise of the ROI. Keeping only the vertical component of the movement trajectories, a fifth order Butterworth filter with a passband of [0.75−5]Hz was applied to remove noise. Balakrishnan et al. followed with PCA decomposition and signal selection to extract an eigenvector that best represents the pulse signal. Using this PCA component signal a heart-rate was estimated using peak detection [18].

Shan et al. built on the previous work of Bal et al. [18] which used head movement to estimate the subjects heart hate. Shan et al. selects the region of interest as an area above the eyes (to avoid any affects caused by facial expressions). Within this ROI a single points was selected as a feature point by using the Shi-Tomasi corner detector function in OpenCV. This function selected a feature point that has the largest eigenvalue for ease in tracking [21]. Shan et al. tested four combinations of filter and analysis consisting of two different filters, a 128 point hamming FIR filter

Table 1 Performance of previous research method using the MAHNOB-HCI DATABASE

Study	Analysis method	MAE (BPM)
Poh et al. [15]	PPG	19.54
Poh et al. [16]	PPG	11.87
Feng et al. [17]	RIPPG	8.05
Shan et al. [21]	BCG	7.88
Bal et al. [18]	BCG	21.68
Hassan et al. [22]	BCG	4.34
Lee et al. [23]	BCG	5.99

and an eighth order Butterworth IIR filter, with both ICA and PCA. This comparison used fifty feature points for each combination as opposed to using a single feature point.

2.3 Results from Previous Work

No previous research was found which directly related to the research methods proposed in this project. There are similar areas of research in non-medical grade heartrate extraction such as PPG and BCG which share a similar processing pipeline. There were three fundamental steps that occurred frequently across the literature review which are as follows systematically.

1. ROI selection and feature point pixel tracking
2. Bandpass filtering of noise
3. Peak frequency analysis

Table 1 contains the predicted heartrate mean absolute error values found in previous research that had been validated against the MAHNOB-HCI DATABASE. A lack of ability to access the MAHNOB-HCI databse prevented direct comparison of the estimated heart rate MAE values in this project to previous research. Despite this issue they are a useful guideline for a general performance validation of the proposed method.

3 Methodology

3.1 Data Collection

The data was collected in the form of a video approximately 3–5 minutes long. The data was sampled from five different people aged from 22–25 years of age. Each person, to the best of their knowledge, was classed as healthy such that there is no reason to consider that their heart rate should be considered abnormal. Alongside the video each subject was asked to wear a health smartwatch to monitor the true (accepted) heart rate during the video; this smart watch uses a built-in optical heart rate sensor. This optical heart rate sensor calculates the beats per minute (BPM) by using PPG analysis. PPG analysis is a widely studied form of optical heart-rate estimation by using the volumetric changes in blood flow in the cardiac cycle.

The smart watch recorded the heart-rate approximately every 2–5 s. This accepted heart-rate reading will be used as a baseline to compare to the predicted heart-rate which is estimated by using the proposed method in this paper. The optical heart rate sensor on the watch does not achieve 100% accuracy however it is a an acceptable baseline to compare with the predicted heart-rate; this comparison can take the form of average value comparison (within an accepted range) and trend comparison (whether the predicted heart-rate can follow the dips and peaks in the heart-rate variability over a set time period).

Three different camera resolutions and frame rates were considered whilst collecting the data, 1920×1080 px at 25 FPS, 2704×1520 px at ≈ 30 FPS and 3840×2160 px at ≈ 15 FPS. This was introduced to better understand the limitations and requirements for the proposed method. The higher the frame-rate and resolution the more expensive the equipment will be however these higher frame rates and resolution may prove to be beneficial to the models performance. Being able to minimise the hardware requirements without diminishing the performance (validity of the signal output) will be more cost-effective for real world applications. This means attempting to identfy a potential cut-off, that is a certain point at which increasing the cameras capability's does not aid in the models performance.

Additional parameters were explored such as general body movement referred to as 'Motion' and recent activity referred to as 'State'. To explore 'Motion' in each video the subject was asked to remain still or move naturally.

These parameters were introduced to better understand the limitations of the proposed method; whether or not additional measures need to be introduced to capture the heart-rate.

The introduction of 'Motion' is to explore the effects that general body movement have whilst trying to estimate a heart-rate. The pixel vibration caused by a heart-beat will be substantially smaller than the pixel displacement caused general body movement. A comparison between several sets of two videos in similar conditions that differ only in the 'Motion' state will determine whether or not general body movement can obscure any existing heart-rate signal.

The 'State' parameter will help identify if the proposed method could capture a higher heart-rate; post exercise the heart-rate is unsettled and prone to fluctuation in the higher frequency range (> 1.67 Hz). It was important to consider the impact the 'State' could have in trying to extract accurate heart-rate predictions.

A full table showing the parameters are shown in Table 2.

Table 2 Name of each video alongside its key components

Video	Length (s)	Framerate (FPS)	Resolution (px)	True HR (BPM)	State	Motion
6946	239.3	29.97	2704 × 1520	68–78(74)	Rest	Still
1652	218.8	25	1920 × 1080	70–81(76)	Rest	Still
1655	27.0	25	1920 × 1080	70–81(76)	rest	Still
1656	202.6	25	1920 × 1080	70–81(76)	Rest	Still
T1	201.7	29.97	2704 × 1520	59–63(60)	Rest	Still
J1	160.0	29.97	2704 × 1520	65–78(74)	Rest	Still
H1	183.0	29.97	2704 × 1520	61–72(66)	Rest	Still
v1	208.3	29.97	2704 × 1520	80–91(85.91)	Rest	Still
v2	208.3	29.97	2704 × 1520	81–93(88.31)	Rest	Natural
v3	213.1	15	3840 × 2160	78–90(85.52)	Rest	Still
v4	236.2	14.9850	3840 × 2160	83–96(89.46)	Rest	Natural
v5	213.9	14.9850	3840 × 2160	80–94(86.22)	Rest	Still
v6	193.3	14.9850	3840 × 2160	107–133(120.03)	Exercise	Still
v7	209.3	29.9700	2704 × 1520	101–109(104.41)	Exercise	Still
v8	304.1	29.9700	2704 × 1520	75–88(82.22)	Rest	Natural
v9	244.1	29.9700	2704 × 1520	74–85(77.96)	Rest	Still
v10	198.0	14.9850	3840 × 2160	74–85(77.96)	Rest	Still
v11	233.4	14.9850	3840 × 2160	72–94(88.62)	Rest	Natural
v12	198.6	14.9850	3840 × 2160	77–91(84.00)	Rest	Natural

Length refers to the length of the video in seconds alongside the resolution and framerate of the video. The true heart-rate was measured with a health smart watch and shows the range of the heart-rate detected. The value shown in '()' is the mean heart-rate throughout the video

3.2 ROI Selection and Tracking

3.2.1 Regions of Interest

A heart beat is result of electrical activity which spreads through the walls of the atria causing them to contract. This motion is noticeable in the lower chest region therefore the lower chest region was selected as a region of interest (refereed to as the 'Heart beat region'). A second region was also selected (referred to as the 'Empty region') which contains a stationary background. This 'Empty' region was to explore how the capability of the tracking algorithm (explained in Sect. 3.3) at its core by detecting a stationary object. This ROI was input manually as a a rectangle of width and height.

Figure 1 represents the region used to extract a heartbeat. This region was common to all of the datasets explored and was selected since heartbeat pulses were expected to be most prominent in the central-left region of the torso which will lead to prominent amplitudes in the FFT output in the heartbeat range.

Fig. 1 Representation of the 'heart beat' region taken in each video. Each white plus represents a pixel that was tracked. This figure corresponds to dataset 4946

Fig. 2 Representation of the 'empty' region taken in each video. Each white plus represents a pixel that was tracked. This figure corresponds to dataset 4946

3.2.2 Empty Region

Figure 2 shows the area selected to monitor background noise. This noise was the result of light variation, slight camera movement

3.3 Implementation of Tracking Algorithm

The pixels were tracked using PointTracker from the Computer Vision Toolbox in MATLAB. The feature point are selected in the region of interest by using the minimum eigenvalue algorithm developed by Shi and Tomasi to detect corner point object in a two dimensional grayscale image [6].

To track the feature points across each video frame the Kanade-Lucas-Tomasi (KLT) algorithm was implemented. Tracking 'n' number of pixels will result in 'n' number of pixel displacement sets.

Previous research by Davis et al. recreated audio through monitoring pixel vibration [24]. This core idea was applied by monitoring the movement of pixels across each frame. As the heart beats it will cause small fluctuations in the clothing surrounding the area. This displacement was determined by tracking the position of each pixel in each frame using the KLT algorithm. The displacement can be expressed as

$$d_i = \sqrt{(x_i - x_{i-1})^2 + (y_i - y_{i-1})^2}, \tag{1}$$

where d_i is the displacement of the pixel at frame i and x_i, y_i are the corresponding coordinates of each pixel at frame i.

There were some limitations to the tracking algorithm used, it required the tracked object to exhibit visual texture on the specified area. This was problematic when attempting to track pixels on clothing that had one colour and some pixels lost the ability to be tracked. To account for this issue coloured tape was applied to the area monitored for ease in tracking. Additional measures used to select appropriate pixels included testing each pixel displacement set and if > 10% of the displacement values were zero then this pixel was removed and not used in further analysis. This method removed any pixels that lost tracking as this could effect further analysis in generating a heartbeat waveform.

The average human heart rate lies within 60–100 beats per minute (BPM) which correspond to frequencies of [1−1.67]Hz using

$$BPM = f \times 60 \tag{2}$$

where f is the frequency of the heartbeat in Hz and 60 refers to 60 s. To determine if a heart beat is visible using a specific pixel the displacement sets needed to be transformed into the frequency domain and analysed; this analysis searched for a prominent peak in the frequency range of [1 − 1.67]Hz.

One approach to estimate the heart rate uses the Fast Fourier Transform (FFT). The FFT algorithm is a faster, more computationally efficient implementation of the Discrete Fourier Transform (DFT) which establishes a relationship between data in the time domain and frequency domain by decomposing a set of values into frequency components. This makes it a suitable method for identifying a heart rate frequency estimate. The discrete Fourier transform of a time series is given by

$$A_k = \sum_{j=0}^{n-1} \exp(-i\frac{2\pi}{n}kj)d_j \qquad (3)$$

where n is the length of the pixel displacement sequence d and $\exp(-i\frac{2\pi}{n})$ is referred to as the nth roots of unity for $k = 0, ..., n - 1$ [25].

The FFT of the pixel displacement distribution decomposes the displacement distributions into a Fourier series A_k of sin waves of different frequencies. Each series is composed of a set of complex number of the from $A_k = B_k \pm iC_k$ which relates to a certain frequency. B is a real number that gives information of the amplitude of the frequency and the imaginary part iC represents the phase of the frequency.

The Fast Fourier transform can only determine frequencies up to the Nyquist frequency which is defined as

$$|f| = \frac{1}{2} \times f_s \qquad (4)$$

where f is the maximum frequency (in Hz) that the FFT can obtain and f_s is the sampling frequency (frames per second) of the video [26]. The heart-rate precision at which the FFT algorithm can determine is dependent of the frequency resolution of the FFT. This relates to the frequency spacing of each bin and is defined as

$$\Delta f_b = \frac{f_s}{N} = \frac{1}{T} \qquad (5)$$

where Δf_b is the frequency spacing of each bin, N is the number of samples in the waveform and T is equal to the time period of the sample [26]. Therefore the time period of each dataset is an import factor to consider to obtain an accurate frequency distribution. The average length of each video sample is approximately two hundred seconds which gives a frequency resolution of $\frac{1}{200} = 5 \times 10^{-3}$Hz. This will give an uncertainty of 2.5×10^{-3} Hz which is the equivalent heart-rate of 0.15 beats per minute. The length of these sample videos should give a sufficiently small frequency bin.

One approach to overcome low frequency spacing is to zero-pad the dataset to increase its length however this will influence and dilute the FFT power spectrum therefore this was not considered.

Once the Fourier distribution had been established the frequency range that correspond to the average human heartbeat were analysed to determine whether there are distinct peaks in the amplitude spectrum. The amplitude spectrum was obtained by using

$$X_k = \sqrt{(B_k)^2 + (iC_k)^2}, \qquad (6)$$

where X_k is the amplitude of the Fourier transform at frame k.

If distinct peaks existed in the amplitude spectrum for frequencies between $[1-1.67]$Hz then this was indicative that a heart rate signal could be extracted from a video.

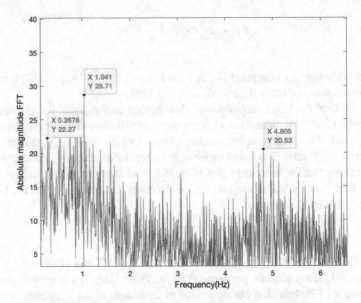

Fig. 3 Amplitude spectrum from the FFT output for a single pixel in the frequency domain. The peak annotated at 1.01 Hz corresponds to a heartbeat value of approximately 60BPM. Other clear peaks can be seen at 0.27 Hz and a wide range at 4.8 Hz which could correspond to respiratory and full body vibrations respectively

Figure 3 demonstrates the FFT output for dataset T1; this shows a strong peak that occurs at 1.01 Hz. The frequency corresponds to a heart rate of ≈ 60BPM. This heart-rate is as expected in comparison to the accepted heart-rate in Table 2.

3.4 Validity of Obtained Frequencies

The frequencies shown in Fig. 3 show promising results but an important step is to ensure the peaks are measurements of the heart rate and not the result of background noise. To accomplish this the pixel tracking was situated on a stationary background area; this area was either a wall or a sofa and known as the 'empty' region in comparison to the original area known as the 'heart beat' region. The expectation was that the FFT output of the 'empty' region would produce near identical amplitudes across the frequency spectrum for noise (noise is further explained in Sect. 3.5) and that these amplitude were significantly smaller that in Fig. 3.

If the FFT output of the 'empty' region was indifferent from the FFT output of the 'heart beat' region then there was no conclusive evidence that the heart rate could be obtained. On the contrary if there was a strong difference, such that significant peaks were present in the 'heart beat' region (within the frequency range of $[1-1.67]$Hz), then that was potentially evidence of a heart beating.

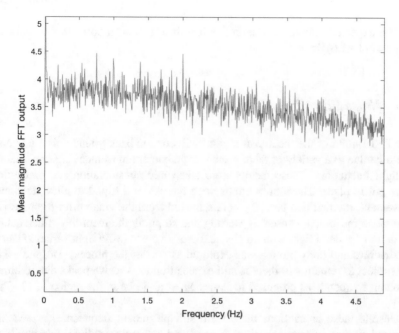

Fig. 4 FFT output on the 'empty' region, this section presents the noise that is present in the signal

Figure 4 shows the FFT output on the 'empty' region; this region shows no significant amplitude spikes across the frequency domain. All frequencies present produced amplitudes in the range of (3.5–4.5) for frequencies in the range [0 − 4.5]Hz. The amplitude spike shown in Fig. 3 shows a distinct peak which occurred at ≈ 1Hz with an amplitude of ≈ 30.

To further compare the amplitudes the FFT output of the 'heart beat' region was compared to the FFT output of the 'empty' region using spectral filtering. This spectral filtering operated within the average heart-rate range of [1−1.67]Hz, with adjustments for datasets that contained a higher heart-rate, to compute the average amplitude $\mu(k)$ is defined by

$$\mu(k) = \mathbb{E}|N(k)|$$

where $N(K)$ is the Fourier transform of the signal in the defined frequency range of the 'empty' region.

$\mu(k)$ was compared to the amplitude spectrum of the 'Heart-beat' region as shown by

$$S(k) =|X(k)| - \mu(k) \tag{7}$$

$$X(k) = \begin{cases} 0, & \text{if } S(k) < \mu(k) \\ X(k) - \mathbb{E}N(k), & \text{otherwise.} \end{cases} \tag{8}$$

The expected output for a definitive heartbeat is a clear spike in the heart-beat range of $(1-1.6)$Hz post filtering.

3.5 Noise Filtration

The FFT output of the 'heartbeat region' will contain background noise; this noise could be due to a variety of factors such slight camera movement, motion artefacts or light fluctuation. These factors were taken into consideration and precautions were put in place. These precautions were the use of a tripod to alleviate camera movement; the first and last 10 s were removed from the video prior to processing to remove movement caused by starting and stopping the recording. Each subject sat directly under a light with no physical crossovers to avoid light variation caused by shadows and body motion was explored as part of the process. Despite asking the subject to remain still there is still expectation to perceive peaks due to human resonant frequency as proposed in James M. et al. to be in the region of $[3-7]$Hz [27]

Despite these precautions there were still prominent frequencies present that required further filtering; respiration produced a large amplitude, this amplitude exists in the low frequency range of $[0.2-0.27]$Hz which dominates the heart-rate amplitudes.

A bandpass filter with a Kaiser window as shown in Fig. 6 was implemented to attenuate any frequencies that lay outside of the average heart-rate at rest. A bandpass filter is a technique used to screen out frequencies that lie out side of the passband range $[F_L, F_U]$ where F_L is the lower limit frequency and F_U is the upper limit frequency.

The average heart-rate at rest for a healthy subject is usually between $[1-1.67]$Hz so this was a useful guideline to initialise the passband range. This frequency range was adjusted to account for the cases where the heart-rate was not at rest and lay above to high cutoff frequency of 1.67Hz. The adjusted frequency range for non resting heart-rates was approximately $[1.5-2.0]$Hz. The Kaiswin window contains a bandwidth of approximately 0.6 Hz.

There are two fundamental types of filters, one being finite impulse response (FIR) and the other is infinite impulse response (IIR) however previous studies have found FIR filters to be more stable and preferable in noise removal [28].

This process utilised a 'Kaiser–Bessel window' based (FIR) filter which is given by

$$\omega_j(n) = \begin{cases} \dfrac{I_0\left[\alpha\sqrt{1-\left[\frac{2n}{N-1}\right]^2}\right]}{I_0(\alpha)}, & \text{for } |n| \leq \frac{N-1}{2} \\ 0, & \text{otherwise} \end{cases} \tag{9}$$

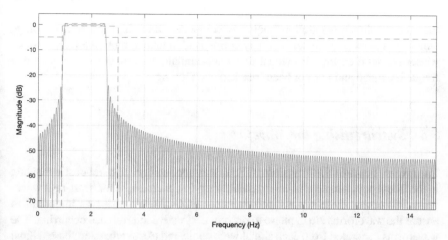

Fig. 5 Kaiser–Bessel window filter between [1–3]Hz

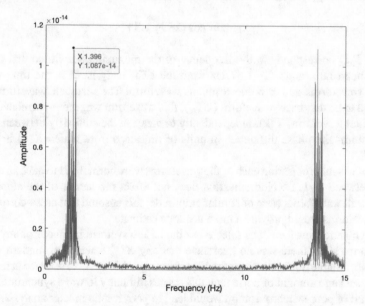

Fig. 6 Average FFT output across every pixel after the Kaiswin passband filtering was implemented

where α is a non-negative real parameter than can be adjusted, α determines the shape of the window. I_0 is the zeroth-order modified Bessel function of the first kind and $N - 1$ is the window length [29] (Fig. 5).

Figure 6 visualises the effect of the Kaiser-Bessel FIR filter with a passband frequency range of $[1-1.67]$Hz. Figure 6 shows that the Kaiser-Bessel filter is effective at alleviating noise outside of the passband frequency range. Frequencies < 1Hz and frequencies > 1.67 Hz were alleviated effectively which removed the issue with the high amplitude respiration frequency component a the higher frequency range which contained frequencies that may correspond to human motion artefacts. After Kaiser-Bessel filtering it is evident that a substantial peak remains in the heart rate frequency range and has not been 'masked' by filtering.

3.6 Synchronising the Waveforms

The inverse Fast Fourier Transform (IFFT) was applied to each pixel post-denoising in the frequency domain and bandpass filtering, to generate a set of heart-beat waveforms. To covert these signals into one final ECG-like signal it was important to ensure the waveforms are in phase to prevent destructive interference occurring. The waveforms generated from each signal were compared to a synthesised 'base' signal waveform ($\mathbf{x}_{base}$) generated through the equation

$$\mathbf{x}_{base} = \cos\left(2\pi F_{max} \mathbf{T}\right), \tag{10}$$

where F_{max} corresponds to the frequency of the maximum amplitude that occurs in the heart rate range $(1-1.67)$Hz from the FFT analysis, T is the time vector of the waveforms and x is the resultant waveform. The resultant waveform was selected to be the 'base' waveform ($\mathbf{x}_{base}$); this waveform was cross-correlated with each pixel waveform, $X(k)$, independently to measure the similarity between them. This returns the phase difference in units of timesteps from which $X(k)$ is phase shifted.

Prior to synchronisation each of the waveforms were normalised to have an amplitude between $(-1, 1)$. Normalisation does not effect the nature of the signal but ensures all waveforms were of similar amplitude; this ensured that cross-correlating the waveforms would provide a more accurate estimate.

A synthetic waveform was selected as the base waveform in favour of any $X(k)$ waveforms since there was no guarantee that any $X(k)$ selected at random would provide a reliable representation of the heartbeat signal. If an $X(k)$ selected at random contains a large amount of noise then this causes difficulty in wave synchronisation, the result of poor synchronisation would lead to poor results in later analysis.

If any pixels exhibited a phase difference significantly greater than the period of one heartbeat then this phase difference is much larger than what was expected. None of the datasets tested had a recorded heart-rate that lay significantly below 60BPM (one beat per-second) which means that a phase-lag of more than one second could be an indicator of destructive interference between the components of the filtered FFT for each pixel. The IFFT reconstructed a heart-beat waveform for each pixel from a superposition of the FFT components with different frequencies and amplitudes which leads to the possibility of destructive interference.

Using the basis that a sinusoidal wave that is $(2n + 1)\pi$ radians out of phase will also be π radians out of phase, any waveform with a phase difference of $\phi > 2\pi$ was reduced by $2\pi N$ such that $\phi < 2\pi$ where N is an integer. A phase differ-

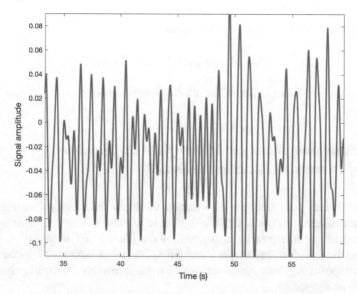

Fig. 7 Visualisation of the heart beat waveform generated by the mean value of each pixel at each timestep

ence of $\phi > 2\pi$ may be a result of including a larger frequency band $(1-1.6)$Hz when filtering thus allowing through a series a components required for a beat phenomena to occur. Introducing a method to shorten this range may prove to be beneficial.

After the waveforms were synchronised, for visualisation purposes, the mean value of each pixel was taken at every timestep. If the waveforms were are synchronised effectively then the expected outcome would be a series of waves with consistent amplitudes which occurred periodically and in phase. Figure 7 demonstrates peaks which occur every second. The amplitude of this wave appears to oscillate and resembles the beat phenomena which occurs from the superposition of similar frequency sine waves.

3.7 Estimating the Beats per Minute Using Peak FFT Analysis

This approach explored the average heart-rate of a subject using the filtered signal in the frequency domain. After noise removal and band-pass filtering in the defined frequency range the expectation is that a prominent average heart-beat frequency peak exists.

This maximum frequency was defined as

$$F_{\max} = \max \left(|X(1)|, ..., |X(n)| \right),$$ (11)

where $|X(n)|$ is the amplitude of the nth frequency component.

3.8 Estimating Beats per Minute from the Number of Prominent Peaks in the Signal

Section 3.7 uses peak frequency estimation to obtain the average heartrate; peak frequency estimation does not give insight as to whether a heart-rate waveform can be generated and interpreted. In this method the filtered FFT output is transferred back into the time domain through inverse-Fast Fourier transform and the number of prominent peaks (n_p) are counted across all n wave-forms sets. The average heart-rate $\overline{HR}$ will be defined as

$$\overline{HR} = \frac{n_p}{T \times n} \times 60$$ (12)

where T is the time period of the video in seconds. If the peak frequency estimation was approximately equal to the accepted average but the prominent peak estimation differed strongly then this was an indication that the generated waveform contained too much noise and further filtering/noise-removal would be required.

3.9 Detecting BPM Through Short-time Fourier Transform

Performing the FFT on a dataset to extract a signal frequency assumes a stationary signal. In practise the BPM is non-stationary and will fluctuate over time which will lead to a range of frequency peaks. One method to approach this issue is the use of the Discrete-time Short-time Fourier transform (STFT) given by

$$X(m, \omega) = \sum_{n=-\inf}^{n=\inf} x(n)\omega_{(n-mL)} \exp^{(-j\omega n)},$$ (13)

where ω_{n-mL} is the shifted sequence, $x(n)$ is the signal input, $\omega(n)$ is the analysis window and L is the window step size. This demonstrates that the window function $\omega(n)$ dictates the portion of the input signal $x(n)$ that is analysed.

This method still has its drawbacks, under consideration of Heisenbergs uncertainty principle selecting a window that was too wide then would prevent instantaneous localisation of time and frequency due to large time uncertainty. This principle

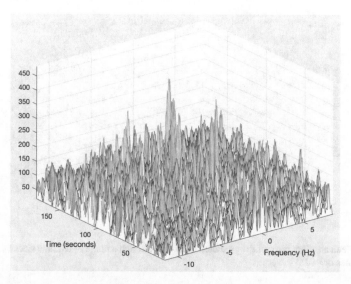

Fig. 8 Visualisation of the Short-Time Fourier Transform on a single pixel. This shows the amplitude of the FFT frequencies across time

could equally be applied to a narrow-window, since this would localise the signal in time however it would also result in a large uncertainty in frequency [30]. Despite these issues this approach will be investigated to determine whether the heart-beat trend can be detected over time through optimising the STFT parameters.

Figure 8 demonstrates how the STFT creates a time-frequency relationship between the pixel displacement values. The 'time' axis corresponds to the STFT window and is plotted alongside the frequency components that are present within this time window. Since there are distinct peaks present within the frequency distribution across the time axis then this approach should allow for a time-series heart rate distribution to be obtained for each dataset.

3.9.1 Optimising the STFT Window Size

The STFT has three primary features that need to be considered, the window choice, window size (w_s) and window overlap size ($w_{overlap}$).

To best optimise the temporal resolution of the STFT output the window size needs to be as small as possible without diminishing the accuracy in the frequency domain. A periodic Hanning window is chosen with the overlap size kept as the default ($w_{overlap} = 0.75 \times w_s$). A large overlap size is important for the STFT to capture all of the significant features within the dataset.

To find the optimal range for the window size the discrete-time STFT was performed on the same dataset with an initial $w_{s=0} = 100$frames with steps of ($\omega_{s+1} = \omega_s + 100$) until ω_s exceeds half the length of the data.

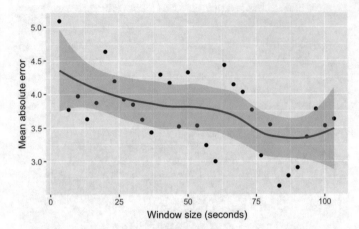

Fig. 9 Mean absolute error of the predicted heart-rate through a range of window sizes of 100 steps (100 frames or ≈ 3 s)

The average error rate will be used to analyse the effectiveness of the window size where the error rate is defined by the mean absolute error (MAE) loss function given by

$$err(P, A) = \frac{\sum_{i=1}^{n} |A_i - P_i|}{n},$$ (14)

where P is the predicted heart rate and A is the true heart rate. MAE was selected over other loss functions such as root mean square error (RMSE) since RMSE is biased towards large errors (larger errors will have more impact than smaller errors) which makes RMSE an unsuitable method for average model performance. MAE is better suited because this is based on the average absolute deviation from the true value and not biased towards small or large errors.

Figure 9 and Fig. 10 show how the error in the predicted heart-rate varies through altering the window size. An important factor to take into consideration is the number of resultant predictions the window size will generate; the higher value window sizes produce a low number of predictions. This makes them less informative for a time series analysis despite offering a lower MAE.

Figures 9 and 10 both show a similar trend, MAE values are high for a smaller window size and level off at a window size of ≈ 35 s. The 35 s corresponds to a window size of 16.7% and 14% of the dataset length for Figs. 9 and 10 respectively.

3.10 Generating ECG Like Signal

A typical ECG signal consists of the P-wave, QRS complexes and T-wave components. The P-wave represents the electrical depolarisation of the atria, the QRS com-

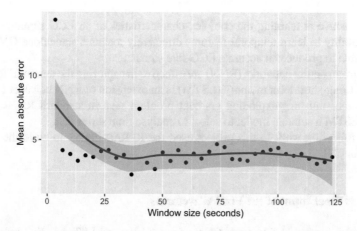

Fig. 10 Mean absolute error of the predicted heart-rate through a range of window sizes of 100 steps (100 frames or ≈ 3 s)

plex shows ventricular depolarisation and the T-wave demonstrates re-polarisation of the ventricles [31].

A collection different heart beat signals from the MIT-BIH Arrhythmia dataset was used in generating an ECG like signal [32]. This datasets has a sampling frequency of 125 Hz and contain 109,446 samples.

The MIT dataset contains ECG signals for a normal heartbeat and heartbeats that are affected by medical conditions such as different arrhythmia's and myocardial infarction. Whilst the collected datasets contain only healthy heartbeats it is important to create a model that will also detect abnormalities as this is critical in any medical applications.

Deep learning models are widely used models and generally outperform traditional machine learning models due to their ability to extract complex features in data. In this section a deep learning model was used to create an ECG like-signal from the heart-beat waveforms.

Two waveforms were used for the creation of an ECG-like signal. The first used the heart-beat waveforms generated through the FFT analysis from Sect. 3.6. The second approach used the heart-beat wavefroms from the STFT output in Sect. 3.9. A small modification to the STFT method was to select the maximum frequency (F_{max}) within the heart-rate range of $[1-1.67]$Hz, with adjustments for cases not at rest, then perform Kaiser-Bessel filtering between $F_{max} \pm 0.15$ Hz. This was to attempt to reduce the amount of noise present in the heart-beat wavefroms. A range of ± 0.15Hz corresponds the heart-rates of ± 9BPM. This gave a heart-rate range of 18BPM for each window size of 35s when performing STFT. When combined with the 80% window overlap this should be sufficient in more robust noise removal tool than the standard FFT analysis.

CNNs are superior in extracting local features in data however they they are not as strong at learning temporal features. The CNNs ability to extract local features may prove effective at learning the complex characteristics of an ECG signal however their inability to learn temporal features effectively means a standalone CNN will not be able to produce an accurate ECG-like signal.

Recurrent neural networks (RNN) are effective in capturing the temporal features of data. Long-short term memory (LSTM) is an extension of RNN however LSTMs address the vanishing/exploding gradient problem seen in standard RNNs which makes LSTM a suitable model to select to analyse time series data.

The following work in this project used the KERAS library to build the neural networks [33].

3.10.1   Direct Input of the Pixel Waveforms

This model combined a 16 layer CNN with an 8 layered LSTM to first extract the local features of the data then the temporal data and was based on the structure used in Abdullah et al. [34]. To compare the effects of a CNN-LSTM hybrid model, each neural network was tested separately with architectures shown in Table 4 for the CNN model and Table 3 for the LSTM model. These were then combined into the CNN-LSTM hybrid with architecture as shown in Table 5.

Using the architectures as shown in Tables 3, 4 and 5 the model was trained on the ECG dataset and predictions were made on the heart-beat wave-forms. The models were expected the learn the characteristic 'QRS' complex of an ECG and replicate this peak predicated on the wave-forms amplitude over time.

Table 3 Structure of the LSTM neural network

Layer number	Name	Additional information
1	Input	–
2	LSTM	50 hidden units with an ReLU activation function
3	Dropout	20% dropout
4	LSTM	50 hidden units with an ReLU activation function
5	Dropout	20% dropout
6	LSTM	50 hidden units with an ReLU activation function
7	Dropout	20% dropout
8	LSTM	50 hidden units with an ReLU activation function
9	Dropout	20% dropout
10	Dense	1 fully connected layer

Table 4 Structure of the 1 dimensional convolutional neural network

Layer number	Name	Additional information
1	Input	–
2	Convolution	Filters = 3. Kernel size = 3. Stride =1. Padding = same
3	Batch normalisation	–
4	ReLU	–
5	Dropout	50% dropout
6	Convolution	Filters = 3. Kernel size = 3. Stride =1. Padding = same
7	Batch normalisation	–
8	ReLU	–
9	Convolution	Filters = 3. Kernel size = 3. Stride =1. Padding = same
10	Batch normalisation	–
11	ReLU	–
12	Dropout	50% dropout
13	Convolution	Filters = 5. Kernel size = 3. Stride =1. Padding = same
14	Batch normalisation	–
15	ReLU	–
16	Global max pooling	–
17	Dense	1 fully conected layer

4 Results

4.1 From FFT Signal Analysis

Table 6 shows the estimated heartrate using the peak amplitude from the FFT output and the number of peaks detected from all of the pixel waveforms. Results from the peak frequency in the FFT analysis were promising for many of the datasets. The table shows the peak frequency obtained in Sect. 3.4 alongside the corresponding heartbeat. The obtained heartbeats, when compared to the mean accepted heartbeat, are close in value and lie within the range so are in excellent agreement. The two exceptions where the predicted values deviate are datasets v6 and v2. The heart-rate lies below the lower end of the range for both datasets. Dataset v2 had 'natural' movement meaning that the subject was not asked to remain still for the entire video but move more freely which may have caused issues in accurately detecting the pixel movement. Dataset v6 measured the heart-rate after exercise in 4k resolution. This heart-rate may cause more inconsistency in beating which leads to more inconsistent pixel displacement. The 4k resolution had maximum frame-rate of 14.975; this frame-rate may not be suitable for measuring erratic pixel movement despite the benefits of a higher resolution; when compared to the other datasets taken at 4k at rest they predict the average heart-rate with a much greater accuracy.

Table 5 Structure of the LSTM-CNN hybrid model used to attempt the generation ECG-like signal

Layer number	Name	Additional information
1	Input	–
2	Convolution	Filters = 3. Kernel size = 3. Stride =1. Padding = same
3	Batch normalisation	–
4	ReLU	–
5	Dropout	50% dropout
6	Convolution	Filters = 3. Kernel size = 3. Stride =1. Padding = same
7	Batch normalisation	–
8	ReLU	–
9	Convolution	Filters = 3. Kernel size = 3. Stride =1. Padding = same
10	Batch normalisation	–
11	ReLU	–
12	Dropout	50% dropout
13	Convolution	Filters = 5. Kernel size = 3. Stride =1. Padding = same
14	Batch normalisation	–
15	ReLU	–
16	LSTM	50 hidden units with an ReLU activation function
17	Dropout	20% dropout
18	LSTM	50 hidden units with an ReLU activation function
19	Dropout	20% dropout
20	LSTM	50 hidden units with an ReLU activation function
21	Dropout	20% dropout
22	LSTM	50 hidden units with an ReLU activation function
23	Dropout	20% dropout
24	Dense	1 fully connected layer

The results from the estimated heart-rate using the number of peaks in the resultant signal were not as accurate as the those from the FFT output, they under-predicted the heart-rate. This could be the result of the passband filtering range allowing destructive interference between the components of the filtered FFT output as discussed in Sect. 3.5. The destructive interference will generate a signal that contains peaks of low amplitude which will not be detected as prominent peaks. This suggests that more work needs to be implemented on filtering noise to produce a signal with a more consistent amplitude with this approach.

Figure 11 shows the absolute error of the predicted mean heart-rate against the accepted heart-rate as shown in Table 6 for each of the movement states alongside the resolution. Table 7 shows the corresponding numerical values. The 'Still' position state indicates that a resolution of 2704×1520 px at 29.97 frames per second exhibits a lower error than a resolution of 3840×2160 px at 15 frames per second. This

Table 6 Table of results showing the obtained average heart rate through analysis of the peak shown from the FFT output and the average number of peaks for the resultant waveform

Dataset	Heart rate from FFT output		Signal output	Accepted heart rate
	Peak frequency (Hz)	Heart rate (BPM)	Heart rate (BPM)	(BPM)
6946	1.195	71.7	69.4	68–78(74)
1652	Na	Na	Na	70–81(76)
1655	Na	Na	Na	70–81(76)
1656	Na	Na	Na	70–81(76)
T1	1.01	60.6	57.3	59–63(60)
J1	1.267	76	71.1	65–78(74)
H1	1.025	61.5	70.5	61–72(66)
v1	1.4756	88.56	72.2	80–91(85.91)
v2	1.25	75.00	77.1	81–93(88.31)
v3	1.418	85.08	71.7	78–90(85.52)
v4	1.396	85.32	–	83–96(89.46)
v5	1.498	89.88	84.92	80–94(86.22)
v6	1.737	104.22	134.91	107–133(120.03)
v7	1.70	102	101.97	101–109(104.41)
v8	1.356	81.36	70.91	75–88(82.22)
v9	1.33	79.80	77.35	71–87(79.46)
v10	1.40	84	75.46	74–85(77.96)
v11	1.358	81.48	50.8	72–94(88.62)
v12	1.467	88.02	78.3	77–91(84.00)

Table 7 Table of values showing the statistic used to plot Fig. 11 alongside an addition MAE value.

	4K still	2K still	4K natural	2K natural
Min	0.44	1.84	4.02	0.860
Lower quartile	2.05	1.84	4.08	0.860
Meadian	3.66	2.15	4.14	7.085
Upper quartile	4.85	2.65	5.64	13.310
Max	6.04	2.65	7.14	13.310
MAE	3.38	2.32	5.1	7.10

All values are measured with units of BPM

suggests that despite the advantage of a higher resolution the disadvantage caused by a lower frame rate is more influential and is important to consider in further work.

The results from datasets 1652, 1655 and 1656 were not tested since the amplitudes of the FFT output on the 'empty' frequencies lay close to the FFT output amplitudes

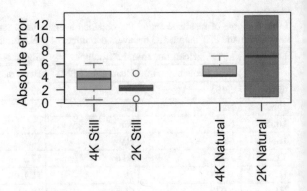

Fig. 11 Absolute error of the predicted heart-rate in BPM for each resolution alongside the movement state

over the 'Heart beat' region which concludes that the frequencies present in the the 'heart-rate' region are indifferent to background noise.

The main difference between datasets 1652, 1655, 1656 and the remaining datasets is the recorded resolution; the framerates are similar with a difference of $\approx$ 5FPS. This result implies that the resolution is crucial to generating and detecting a clear heart rate. To further explore this more data needs to be collected at a higher resolution.

4.2 Short Time Fourier Analysis

The covariance between the the predicted BPM against the accepted BPM was used to quantify the similarity of the time series values, compare the trends and to distinguish if the STFT can detect sudden peaks or drops in the heart-rate. A positive covariance indicates that the the predicted heart rate and accepted heart rate increase and decrease together. Conversely a negative heart-rate indicates the opposite; the predicted heart moves in the opposite direction to the accepted heart rate. A perfect covariance would be a covariance of one as this represents a perfect linear relationship between the accepted and predicted heart-rate.

Table 8 demonstrates the covariance between the predicted heart-rate against the accepted heart-rate, dataset v4 shows the best outcome with a covariance of 0.92 indicating that the STFT on this dataset managed to accurately detect the heart-rate trend. Conversely datasets v6, v8 and v11 all show a negative covariance. Dataset v6 also shows poor performance through direct FFT analysis. Dataset v6 measures the highest accepted heart-rate values (heart rate frequencies which exceed 2 Hz); the frequency region that exceeded 2Hz contains a lot of noise which could influence the final result through masking the heart-beat frequency. In the case for negative covariance it was also beneficial to directly analyse the trends through a graphical representation.

Figure 12 shows the predicted and accepted heart-rates of v11 through time. Direct interpretation of the BPM in Fig. 12 shows that the predicted heart-rate is in the correct range however the peaks for the predicted heart-rate appear to have a positive phase

Table 8 Covariance between the predicted heart-rate values versus the accepted heart-rate values

Dataset	Covariance
v1	0.006696461
v2	–
v3	0.5772765
v4	0.9206335
v5	0.8699091
v6	−8.236898
v7	1.571281
v8	−0.3991068
v9	3.535848
v10	0.5152384
v11	−5.441606

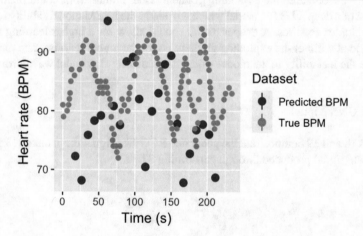

Fig. 12 Predicted heart-rate values alongside the accepted heart-rate values through short-time Fourier transform using an overlap length of 75% and a window size of 35% of the length of the dataset for dataset v11

difference of approximately (40−50)s. The phase difference seen in Fig. 12 could be the result of a time delay between the initiation of the video alongside the start of the heart-rate monitor from the watch; each had to be initialised independently. Furthermore the STFT has uncertainty in the time domain which may be one of the major underlying causes of this lag. Dataset v11 was also recorded at 3840 × 2160 px resolution at approximately 15 frames per second which may be to low of a frame-rate to depict the time resolution to a high enough accuracy.

Figure 13 does not show the same issue as Fig. 12; the peaks between the predicted heart-rate and accepted heart-rate align in the time domain however there are large peaks that appear in the first and last few instances of the predicted values. These large peaks may have been the cause of the large covariance value (3.5) associated with dataset v9.

4.3 Generating ECG Like Signal

4.3.1 LSTM Model

Using the STFT signal output for dataset vid13, Figs. 14 and 15 were constructed
using the LSTM model and shown in Table 3. Figure 14 shows that the model is
capable of learning the spike associated with the QRS complex in an ECG signal at
101 and 102 s. This spike is not consistent across the entire time domain as seen in
Fig. 15. Figure 15 shows the inability of the model to learn the ECG characteristics
as after 76 s the predicted output resembles a sinusoidal signal rather than an ECG
like signal where may be because the learning rate is too low and the LSTM model
is unable to produce the ECG characteristics. This was attempted to be rectified by
a higher learning rate (see Figs. 16, 17); however, any learning rate that exceeded
1×10^{-4} encountered the exploding gradient issue. Future work could include the
creation of a deeper LSTM model which will aid in the LSTM models ability to learn
more complex features. A deeper model should allow for a higher learning rate to
be included without the exploding gradient issue; the higher learning rate may also
remove the instability in the model where its predicting sinusoidal waveforms.

4.3.2 CNN Model

Figures 18 and 19 demonstrate the generated ECG-like signal for dataset v13 for the
heart-rate signal produced through performing STFT.

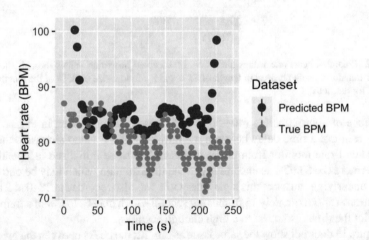

Fig. 13 Predicted heart-rate values alongside the accepted heart-rate values through short-time
Fourier transform using an overlap length of 75% and a window size of 35% of the length of the
dataset for dataset v11

Figures 18, 19, 20 and 21 that the CNN model (shown in Table 4) is more adept at learning the sharp peaks associated with the ECG signal than the LSTM model however the overall signal does not give a clear indication of each heartbeat.

4.3.3 CNN-LSTM Model

Figure 22 shows the generated ECG-like signal for the CNN-LSTM model as shown in Table 5. Figure 22 contains sections which show similar traits to an ECG signal at 77–79 s; there is a drop prior to the peak (similar to the QRS complex) followed by a flattened area (similar to the ST segment of an ECG signal). After 79 s in Fig. 22 the model loses these traits and begins to resemble a sinusoidal signal; this may indicate that the learning rate was too low or the four LSTM layer may not be sufficient for hybrid neural network to learn the complexity of the ECG signal.

Figure 23 demonstrates the effects of a learning rate which is too high, at 110–125 epochs the MAE loss function 'exploded'. This is indication of the loss function attempting to cause weight updates that are too drastic and diverge from the optimum.

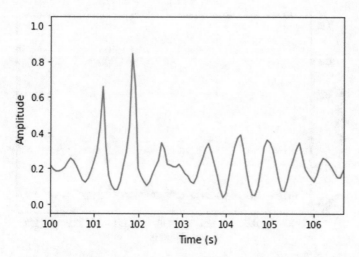

Fig. 14 Generated ECG-like signal using the LSTM model using the 'Adam' optimiser with a learning rate of 1×10^{-6} and a decay of 1×10^{-3}. This ECG-like signal corresponds the the STFT output of dataset v13

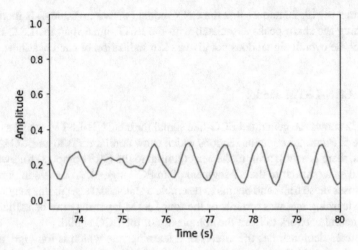

Fig. 15 Generated ECG-like signal using the LSTM model using the 'Adam' optimiser with a learning rate of 1×10^{-6} and a decay of 1×10^{-3}. This ECG-like signal corresponds the the STFT output of dataset v13

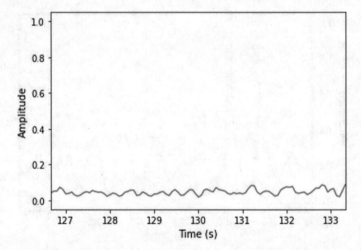

Fig. 16 Generated ECG-like signal using the LSTM model using the 'Adam' optimiser with a learning rate of 1×10^{-4} and a decay of 1×10^{-3}. This ECG-like signal corresponds the the filtered FFT output of dataset v13

5 Conclusion

5.1 Key Findings

The aim of this project was to devise a novel method for extracting a heart rate through remote imaging. Two different approaches were used in an attempt to detect a human

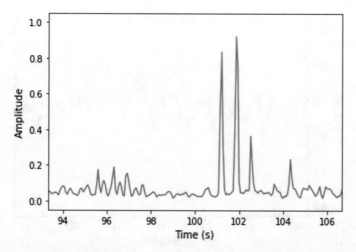

Fig. 17 Generated ECG-like signal using the LSTM model using the 'Adam' optimiser with a learning rate of 1×10^{-4} and a decay of 1×10^{-3}. This ECG-like signal corresponds the the filtered FFT output of dataset v13

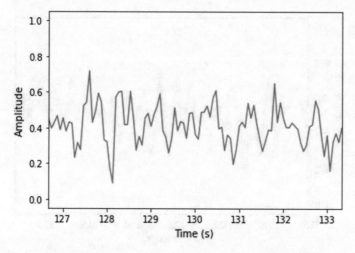

Fig. 18 Generated ECG-like signal using the CNN model using the 'Adam' optimiser with a learning rate of 1×10^{-6} and a decay of 1×10^{-3}. This ECG-like signal corresponds the the STFT output of dataset v13

heart-beat from tracked pixel displacement. The first method implemented the Fast Fourier transform on the pixel displacement values over a period of approximately three minutes.

Peak frequency analysis demonstrated that the Fast Fourier transform was capable of detecting the mean heart-beat frequency close to the accepted mean heart-rate value and lay within the heart-rate range for 74% of the cases tested. This percentage

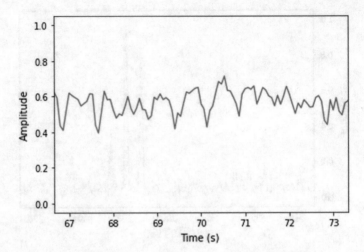

Fig. 19 Generated ECG-like signal using the CNN model using the 'Adam' optimiser with a learning rate of 1×10^{-6} and a decay of 1×10^{-3}. This ECG-like signal corresponds the the STFT output of dataset v13

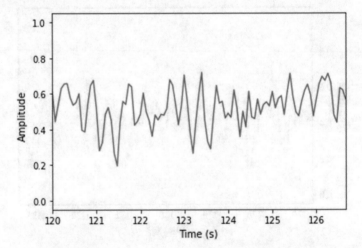

Fig. 20 Generated ECG-like signal using the CNN model using the 'Adam' optimiser with a learning rate of 1×10^{-4} and a decay of 1×10^{-3}. This ECG-like signal corresponds the filtered FFT output of dataset v13

includes the cases which were not tested; all the videos recorded with a resolution of 1920×1080 did not produce a frequency distribution in the 'heart-beat' region that was distinguishable from background noise. Datasets v2 and v6 produced a maximum frequency that underestimated the mean heart-beat and lay marginally below the lower end of the accepted heart-beat range.

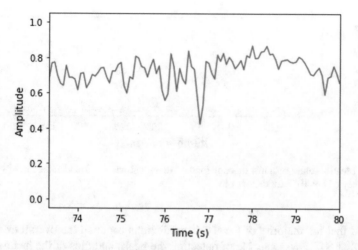

Fig. 21 Generated ECG-like signal using the CNN model using the 'Adam' optimiser with a learning rate of 1×10^{-4} and a decay of 1×10^{-3}. This ECG-like signal corresponds the filtered FFT output of dataset v13

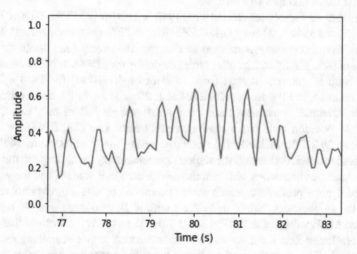

Fig. 22 Generated ECG-like figure using the CNN-LSTM architecture as shown in Table 5. This run using the 'Adam' optimiser with a learning rate of 1×10^{-6} and a decay of 1×10^{-3} over two hundred epochs

The second approach used to estimate the heart-beat was the Short-time Fourier transform; this method estimated the beats per minute over a specified time window. The STFT time window size was optimised by using a MAE loss function to minimise the error on the predicted heart-rate. A window size of 17% of the dataset was found to be the optimal size across two different video with different resolutions and framerate. The covariance between the accepted time series heart-rate and predicted time series heart-rate was used as a method to capture the similarity of the trend. The covariance

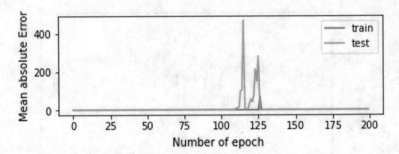

Fig. 23 Mean absolute error loss function for the 'Adam' optimiser with a learning rate of 1×10^{-4} and a decay of 1×10^{-3} for dataset v13

revealed that the majority of the STFT predictions captured the overall trend; this means that STFT was capable of detecting the peaks and dips of the heart-rate. In a minority of cases the predicted values exhibited the peaks and dips of the varying heart-rate as seen through visual analysis with a time lag which may be a result of poor time resolution in STFT analysis.

Three different resolutions tested were 1980×1080 px (25 FPS), 2704×1520 px (29.7 FPS) and 3840×2160 px (14.95 FPS) for two different movement states. Using the peak frequency analysis method to estimate the average heart-rate the 'Still' movement state produced a smaller error (median error of 3.66BPM) in comparison to the 'Natural' movement state (median error of 4.14BPM) for 3840×2160 px resolution at 14.95 FPS. Results for 2704×1520 px at 29.7 FPS were inconclusive since the 'Natural' movement state has a small sample size of only two datasets. The 'Still' position state has a smaller median error size (2.15BPM) for 2704×1520 px at 29.7 FPS (2.15BPM) in comparison to 3840×2160 px at 14.95 FPS. From these results it is clear that the subjects movement negatively impacts the models performance. A comparison between the generated MAE values from the predicted heartrate against previous research shows promising results. For datasets recorded where the subject was 'Still'; the videos with a 2k resolution and 4k resolution produced MAE values of 2.32BPM and 3.38BPM respectively. Both of these MAE values are lower than those found in literature which shows promising results for this method of heartrate extraction. When the subject was recorded with 'Natural' movement the MAE values were higher and calculated to be 7.10BPM and 5.1BPM for video resolutions of 2k and 4k respectively. Both of these MAE values are smaller than over half of the MAE values found in previous research.

These results imply that a heart-rate can be tracked through use of FFT, the heart-rate can be detected with videos sampled with a resolution of 3840×2160 px at 14.95 FPS however it is not recommend due to the higher error rate. A minimum of 30 frames per second is recommended to gain more precise readings. This was the minimum recommended value in Blackford et al. with non-contact PPG heart-rate analysis [13].

Two different neural network structures were used in an attempt to create an ECG-like signal for the FFT heart-beat waveform and the STFT heart-beat waveform.

The singular neural network used was a deep network CNN-LSTM hybrid model. For the STFT heart-beat waveforms, this model was able to learn the QRS complex however it was not stable throughout the entire time period. This instability may be caused by destructive interference of the heart-beat waveforms or noise that is still present in the signal. Based on the ECG results from this model it can not be concluded that the final aim 'Can an ECG-like signal be generated using the heart-rate waveforms?' was achieved. The ECG-like signals were not consistent throughout the entire domain however it may be possible with further research.

5.2 Limitations

Higher frame-rates and resolutions were not able to be tested due to limitations with the camera used but future work would recommend testing with higher frame-rate equipment.

The algorithm used to select the feature points has the option to keep a predefined number of strongest corner points. The number of feature points should be carefully selected to best optimise the pixel tracking.

5.3 Future Research

The MAE values in comparison to previous research showed promising results however the heartrate extraction method in this paper used a narrow bandwidth bandpass filter to extract either the low frequency heart range or the higher frequency heart range. The narrow bandwidth limits the proposed method for certain scenarios in regards to the subjects recent activity. Further research could test the affects over a larger bandpass bandwidth to test the limitations of the proposed method.

Additional work could go into removing pixels which may not have accurately tracked the shirt fluctuation; each pixel was assumed to have perfect fluctuation. Not every pixel may be able to represent a heart-rate; pixels that are tracked over areas of large creases may not fluctuate when the heart-beats. Eliminating the pixels which did not sufficiently track the heartbeat would reduce noise and aid in estimating a more accurate heart rate.

Potentially, our non-invasive heart monitoring method can serve as a remote healthcare surveillance tool for smart homecare. We will investigate the application of our innovative method on this area in our future work.

References

1. Shim H, Lee JH, Hwang SO, Yoon HR, Yoon YR (2009) Development of heart rate monitoring for mobile telemedicine using smartphone. In: 13th international conference on biomedical engineering. Springer, Berlin, pp 1116–1119
2. Castaneda D, Esparza A, Ghamari M, Soltanpur C, Nazeran H (2018) A review on wearable photoplethysmography sensors and their potential future applications in health care. Int J Biosens Bioelectron 4(4):195
3. Kwon S, Kim J, Lee D, Park K (2015) Roi analysis for remote photoplethysmography on facial video. In: 37th annual international conference of the IEEE engineering in medicine and biology society (EMBC). IEEE, pp 4938–4941
4. Duran G, Tapiero I, Michael GA (2018) Resting heart rate: a physiological predicator of lie detection ability. Physiol Behav 186:10–15
5. Zhu J, Ji L, Liu C (2019) Heart rate variability monitoring for emotion and disorders of emotion. Physiol. Meas. 40(6):064004
6. Shi J (1994) Good features to track. In: Proceedings of IEEE conference on computer vision and pattern recognition. IEEE 593–600
7. Sani NM, Mansor W, Lee KY, Zainudin NA, Mahrim S (2015) Determination of heart rate from photoplethysmogram using fast fourier transform. In: 2015 international conference on BioSignal analysis, processing and systems (ICBAPS). IEEE, pp 168–170
8. Aisuwarya R, Hendrick H, Meitiza M (2019) Analysis of cardiac frequency on photoplethysmograph (ppg) synthesis for detecting heart rate using fast fourier transform (fft). In 2019 international conference on electrical engineering and computer science (ICECOS). IEEE, pp 391–395
9. Asthana A, Zafeiriou S, Cheng S, Pantic M (2013) Robust discriminative response map fitting with constrained local models. In Proceedings of the IEEE conference on computer vision and pattern recognition, pp 3444–3451
10. Tomasi C, Kanade T (1991) Detection and tracking of point. Int J Comput Vis 9:137–154
11. Li X, Chen J, Zhao G, Pietikainen M (2014) Remote heart rate measurement from face videos under realistic situations. In: Proceedings of the IEEE conference on computer vision and pattern recognition, pp 4264–4271
12. Bush I (2016) Measuring heart rate from video. In: Standford Computer Science, retrieved at: https://web.stanford.edu/class/cs231a/prev_projects_2016/finalReport.pdf
13. Blackford EB, Estepp JR (2015) Effects of frame rate and image resolution on pulse rate measured using multiple camera imaging photoplethysmography. In: Medical imaging 2015: biomedical applications in molecular, structural, and functional imaging, vol 9417. International Society for Optics and Photonics, p 94172D
14. Wang C, Pun T, Chanel G (2018) A comparative survey of methods for remote heart rate detection from frontal face videos. Front Bioeng Biotechnol 6:33
15. Poh M-Z, McDuff DJ, Picard RW (2010) Non-contact, automated cardiac pulse measurements using video imaging and blind source separation. Opt Express 18(10):10,762–10,774
16. Poh M-Z, McDuff DJ, Picard RW (2010) Advancements in noncontact, multiparameter physiological measurements using a webcam. IEEE Trans Biomed Eng 58(1):7–11
17. Feng L, Po L-M, Xu X, Li Y, Ma R (2014) Motion-resistant remote imaging photoplethysmography based on the optical properties of skin. IEEE Trans Circ Syst Video Technol 25(5):879–891 (064004)
18. Balakrishnan G, Durand F, Guttag J (2013) Detecting pulse from head motions in video. In: Proceedings of the IEEE conference on computer vision and pattern recognition, pp 3430–3437
19. Viola P, Jones M (2001) Rapid object detection using a boosted cascade of simple features. In: Proceedings of the 2001 IEEE computer society conference on computer vision and pattern recognition. CVPR 2001, vol 1. IEEE, pp I–I
20. Bradski G (2000) The openCV library. Dr. Dobb's J Softw Tools 120:122–125
21. Shan L, Yu M (2013) Video-based heart rate measurement using head motion tracking and ICA. In: 6th international congress on image and signal processing (CISP), vol 1. IEEE 160–164

22. Hassan MA, Malik AS, Fofi D, Saad NM, Ali YS, Meriaudeau F (2017) Video-based heartbeat rate measuring method using ballistocardiography. IEEE Sens J 17(14):4544–4557
23. Lee H, Cho A, Lee S, Whang M (2019) Vision-based measurement of heart rate from ballistocardiographic head movements using unsupervised clustering. Sensors 19(15):3263
24. Davis A, Rubinstein M, Wadhwa N, Mysore GJ, Durand F, Freeman WT (2014) The visual microphone: passive recovery of sound from video. ACM Trans Graph 33(4):1–10. https://doi.org/10.1145/2601097.2601119
25. Heckbert P (1995) Fourier transforms and the fast Fourier transform (FFT) algorithm. Comput Graph 2:15–463
26. Lévesque L (2014) Nyquist sampling theorem: understanding the illusion of a spinning wheel captured with a video camera. Phys Educ 49(6):697
27. Brownjohn JM, Zheng X (2001) Discussion of human resonant frequency. In: Second international conference on experimental mechanics, vol 4317. International Society for Optics and Photonics, pp 469–474
28. Kumar KS, Yazdanpanah B, Kumar PR (2015) Removal of noise from electrocardiogram using digital FIR and IIR filters with various methods. In International conference on communications and signal processing (ICCSP). IEEE 0157–0162
29. Chavan MS, Agarwala R, Uplane M (2006) Use of Kaiser window for ECG processing. In: Proceedings of the 5th WSEAS international conference on signal processing, robotics and automation, Madrid, Spain
30. Mazurova E, Lapshin A (2013) On the action of Heisenberg's uncertainty principle in discrete linear methods for calculating the components of the deflection of the vertical. In: EGU general assembly conference abstracts, pp EGU2013–2466
31. Prasad ST, Varadarajan S (2014) ECG signal analysis: different approaches. Int J Eng Trends Technol 7(5):212–216
32. Moody GB, Mark RG (2001) The impact of the MIT-BIH arrhythmia database. IEEE Eng Med Biol Mag 20(3):45–50
33. Chollet F et al (2015) Keras. https://keras.io
34. Abdullah LA, Al-Ani MS (2020) CNN-LSTM based model for ECG arrhythmias and myocardial infarction classification. Adv Sci Technol Eng Syst J 5(5):601–606

A Survey on the Integration of Blockchain and IoT: Challenges and Opportunities

Mwrwan Abubakar, Zakwan Jarocheh, Ahmed Al-Dubai, and Xiaodong Liu

Abstract Since Satoshi Nakamoto first introduced the blockchain as an open-source project for secure financial transactions, it has attracted the scientific community's interest, paving the way for addressing problems in domains other than cryptocurrencies, one of them being the Internet of Things (IoT). However, to demonstrate this potential, a clear understanding of blockchain technology and its suitability to meet the IoT's underlying security requirements is required. To accomplish this goal, this study intends to provide a coherent and comprehensive survey on blockchain integration with IoT identifying the limitations and benefits of Blockchain in IoT applications. The survey presented an overview of blockchain and IoT and illustrated their principal architecture. The state-of-the-art blockchain efforts in different IoT domains are also reviewed. Then presents the current challenges of integrating blockchain and IoT and looks at the recently proposed solutions. Finally, the potential future research directions for integrating blockchain and IoT is discussed. Based on the study's findings, it is hoped that this survey will serve as a reference and source of motivation for future research directions in this area.

Keywords Blockchain technology · IoT · Survey · Integration of blockchain and IoT

M. Abubakar (✉) · Z. Jarocheh · A. Al-Dubai · X. Liu
Edinburgh Napier University, Edinburgh, UK
e-mail: m.abubakar@napier.ac.uk

Z. Jarocheh
e-mail: z.jaroucheh@napier.ac.uk

A. Al-Dubai
e-mail: a.al-dubai@napier.ac.uk

X. Liu
e-mail: x.liu@napier.ac.uk

© Springer Nature Switzerland AG 2022
R. Jiang et al. (eds.), *Big Data Privacy and Security in Smart Cities*,
Advanced Sciences and Technologies for Security Applications,
https://doi.org/10.1007/978-3-031-04424-3_11

1   Introduction

The Internet of Things (IoT) [1] has transformed our lives and represented a significant step forward in how we will use technology. IoT enables real-time sensing capabilities, empowering various areas such as agriculture, healthcare, smart homes and supply chain, etc., thereby contributing to the evolution of new solutions, such as smart cities, to improve citizens' quality of life. However, with the rise in data privacy breaches, IoT adopters must look for secure alternatives rather than depending on a centralised model. On the other hand, blockchain technology has emerged recently as a distributed computing paradigm that successfully overcomes issues related to the trust of a centralised party. Blockchain technology provided a distributed, incorruptible and tamper-resistant ledger that can be operated without the control of an intermediary or any third party, which has been utilised in the first place to secure financial transactions and realising cryptocurrency such as Bitcoin [2] and Ethereum [3]. However, the second generation of blockchain and distributed ledger technologies are more general-purpose. Instead of recording financial transactions, it can record data for any other type of application. Additionally, blockchain can execute and install a script known as a smart contract, such as on the Ethereum Blockchain system, allowing for the expansion of blockchain's applicability to embrace emerging technologies such as the Internet of Things. The decentralised structure of the blockchain can provide secure trust methods where data will be trusted by all nodes in the blockchain network. Thus providing a decentralised, secure ledger to store and then validate transactions in a decentralised manner. The concept of integrating IoT and Blockchain technology has attracted significant scholarly interest in recent years and is viewed as a potential solution for present IoT systems' security concerns. However, blockchain also encounters a number of critical challenges inherent in the IoT, including a large number of IoT devices, a heterogeneous network structure, low computational capacity and limited communication bandwidth. Therefore, to prove this potential, there must be a clear understanding of blockchain technology and its suitability to satisfy the underlying security requirements of IoT. To achieve this goal, in this study, we aim to provide a coherent and comprehensive survey on blockchain integration with IoT.

The remainder of this paper is organised as follows. The paper starts by briefly describing blockchain technology and then reviewing the related literature that surveys the integration of the blockchain with IoT in Sect. 2. Section 3 presents an overview background on IoT and blockchain technology, illustrating their principal architecture, and special emphasis is placed on the unique characteristics of blockchain in IoT applications. The survey further elaborates on and presents the state-of-the-art efforts in different IoT domains in Sect. 4. In Sect. 5, the paper presents the current challenges and issues of integrating the blockchain and IoT, understanding their impacts on the scalable design of Blockchain for IoT applications and then reviewing the currently proposed solutions. After that, the paper discusses the possible future research directions in integrating the blockchain with IoT in Sect. 6. Finally, concluding the paper in Sect. 7.

2 Literature Review

This section reviews the literature and summarises the current surveys of the blockchain implementation in IoT. Recently, the possibility of integrating IoT and Blockchain technology has sparked considerable interest in the research community. Numerous efforts have been made to review current integrating blockchain and IoT initiatives and investigate the benefits and challenges associated with such integration. However, these surveys differ in their depth and scope. For example, the works presented in [4–10] provided broad overviews of blockchain integration with IoT. Alternatively, several other surveys [11–16] examined specific aspects of blockchain implementation for IoT. Table 1 presented to summarise the main aspects covered by the previous surveys and their major contributions.

For instance, authors of [4] covers the state of the research on the Internet of Things and its possible applications, as well as a case study on distributed IoT. Additionally, it examines numerous difficulties and potential associated with decentralised ecosystems and open research topics and future initiatives. However, these survey did not show the current blockchain implementation in different IoT domains. The work presented in Alam [5] discussed the need for a new framework for IoT, which can lead to continued growth in IoT use and popularity. The presented work proposed developing a framework for middleware on the internet of smart devices network using blockchain technology. The survey goes on to discuss blockchain's role in IoT. However, the blockchain's advantages and disadvantages are briefly discussed, and only a brief overview of blockchain implementation in IoT applications is presented. The authors of [6] also investigated the integration of blockchain technology and IoT. They coined the term "Blockchain of Things" (BCoT) to refer to this combination of blockchain and IoT. The paper also analyses the convergence of blockchain and IoT and propose a BCoT architecture. Additionally, it discusses the issues surrounding the use of blockchain in 5G applications as well as industrial BCoT applications. Finally, they discussed the areas of research that remain unexplored in this promising field. Another interesting work is presented in Dai et al. [7], which summarises the significant findings from existing research on IoT challenges.

Table 1 Previous surveys and their major contributions

Contributions	Recent surveys	Our contribution
Comprehensive survey	[7, 8]	✓
Blockchain IoT applications	[10, 14]	✓
Challenges	[5, 7, 10, 14]	✓
Open research directions	[4, 6, 8–10, 13]	✓
Security and privacy issues	[7]	✓
Smart contract for IoT	[16]	✓
Case studies	[4, 11, 12, 15]	✓
Introducing solutions	[7]	✓

The paper discussed the technical characteristics of blockchain technology and the benefits of incorporating it into IoT. Additionally, the paper discussed several security challenges, including key management, intrusion detection, access control, and privacy protection, as well as the technology available to address them. Finally, discussed and summarised the industry's prospects. Similarly, the authors of [8] provide an in-depth investigation of current Blockchain technologies. Then, they determined which Blockchain properties are well-suited for IoT applications. Adaptations and enhancements to the Blockchain consensus protocols and data structures are also discussed. Future research directions for effective Blockchain integration into IoT networks are discussed. This survey does not cover current blockchain applications in different IoT fields. Another survey is presented in Atlam et al. [9], which discusses the integration of IoT with blockchain technology and provides an overview of IoT and blockchain technology. Then, the blockchain as a service for the IoT is demonstrated, demonstrating how various features of blockchain technology can be implemented as a service for a variety of IoT applications. Following that, the impact of integrating artificial intelligence (AI) on both IoT and blockchain is discussed. Although this survey provided useful information, it does not survey the current state of integrating blockchain in IoT. The authors of [10] reviewed recent state-of-the-art advances in Blockchain for IoT, Blockchain for Cloud IoT, and Blockchain for Fog IoT in the context of eHealth, smart cities, and intelligent transportation, among other applications. Additionally, obstacles, research gaps, and potential solutions are discussed. In [11], the authors discuss how blockchain technology will eventually solve the problem of centralised cloud computing. The study presented the use of Blockchain in smart homes through the lens of three tiers and discussed how Blockchain could be used to safeguard data and transactions to provide security on the Internet of smart homes. However, this survey is very strict in a specific scenario. For existing blockchain-based access control, the authors of [12] provided a comprehensive review of existing blockchain-based access control models. For blockchain challenges and solutions, the work presented in Zhu and Badr [13] identifies digital identity-related challenges and solutions for the Internet in general. The authors of [13] discussed digital identity-related challenges and solutions. The paper reviewed current solutions and their ability to address IoT requirements for scalability, interoperability, mobility, security, and privacy. Then looks into the emerging blockchain-based self-sovereign identity solutions and identifies projects and startups addressing IoT identity issues. Furthermore, the survey discusses the difficulties inherent in developing identity management systems for IoT. Finally, it discusses future research direction in this area. Another survey is presented in Chowdhury et al. [14] reviews the current state of the art research in IoT applications. The survey studied the potential cases in different IoT domains and investigated the challenges and solutions of using blockchain to develop secure IoT applications.

The work presented in Hassanien et al. [15] discusses blockchain implementations for smart cities. The article presents an overview of the current state of blockchain technology utilisation in smart cities. The paper reviewed some of the related studies and presented a new blockchain architecture for smart cities. The proposed design makes use of Hyperledger Indy. The use of smart contracts in IoT is reviewed in

Pranto et al. [16]. The study investigated different aspects of using blockchain and smart contracts in securing IoT systems. Then, developed a system in which all connected parties engage via a smart contract. Finally, discussed the strength and weaknesses of the proposed solution.

3 An Overview of IoT and Blockchain

3.1 Internet of Things (IoT)

Kevin Ashton first introduced this term in 1999 [17] as a bind between the Internet and radio frequency identification. Technically, it refers to objects that can be connected to each other over the Internet. These objects could be any device embedded with software, electronics or sensors. These devices are capable of communicating with one another by gathering, sharing, and exchanging data. The global need for IoT app development has never been higher. The cause for this growth is the rising demand for industrial automation, which supports the development of IoT solutions. According to findstack statistic [18], the total number of active IoT devices will likely reach 22 billion by 2025. However, because IoT communication is wireless, the system is more susceptible to message manipulation, message eavesdropping, and identity spoofing.

3.2 IoT Framework

Even though no single IoT architecture is universally agreed upon, the most basic and widely accepted format is a three-layer architecture: Devices, Network, and Applications, as shown in Fig. 1.

Perception Layer The primary function of this layer is to gather meaningful data from objects or the environment. This layer comprises various sensors, ranging from basic temperature sensors to advanced medical and industrial sensors.

Network layer This layer represents the connectivity between the devices layer and the cloud. Secure communication between things and the cloud is one of the most critical requirements. Current security technology that has been widely used is SSL/TLS encryption. However, another issue can arise when implementing cryptography in IoT constrained devices, e.g., using RSA with 1024 bit key in a limited RAM and storage micro-controller.

Application layer This layer is responsible for managing IoT systems by controlling access to users and devices, enforcing rules and policies, coordinating automation across multiple devices, managing users and devices to identify and provide optimal access control based on their privileges, and finally, auditing and monitoring data.

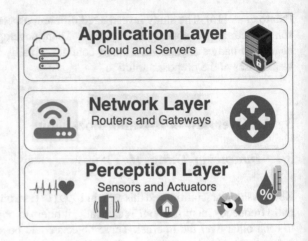

Fig. 1 The three layers of IoT architecture: perception, network, and applications layers

3.3 IoT Security Challenges

Most state-of-the-art IoT infrastructures are heavily centralised, prone to a single point of failure. The entire network infrastructure risks being paralysed in the event of a failure in the centralised servers, which hinder scalability and wide adoption of the IoT. The centralised infrastructure has also raised severe privacy and security concerns. Because the centralised methods for providing privacy, security and data handling necessitate high-end servers that mandates trusting a third party for data handling. In the centralised model, users have limited control over their data. Users are required to trust such entities to handle their personal data, which is prone to the risk of being deleted or tampered with. In addition, the centralised infrastructure leads to higher latency for end-to-end communications and lacks guaranteed accountability and traceability [19].

Resource constrained IoT devices are always resource-constrained, preventing them from implementing effective advanced security solutions. When encryption is used for authentication, for example, certain advanced encryption methods can result in issues such as decreased computing performance, increased hardware power consumption, and so on.

Heterogeneity of IoT systems Another challenge is the heterogeneity of the IoT system's components, as IoT systems might comprise devices from different vendors, each having its own platform and technology. Heterogeneity is seen in various IoT devices, communication protocols, and data formats. The heterogeneity is also the root of other challenges such as interoperability, privacy and security.

Lack of encryption Even when data is encrypted, vulnerabilities may exist if the encryption is incomplete or incorrectly set. Encryption should also be used to secure sensitive data stored on a device (at rest). Typical vulnerabilities include the lack of encryption, which occurs when API tokens or credentials are stored in plain text on a device. Another issue might develop when implementing cryptography in IoT-constrained devices, for example, when employing RSA with a 1024-bit key in a microcontroller with restricted RAM and storage.

Complexity of networks In the Internet of Things, a variety of communication and network protocols coexist. The data collected by IoT devices will be transmitted to server stations or other low-power devices. These devices require a software layer to enable access to hardware functionalities and protocols to connect these devices to other communication protocols [20].

Poor interoperability It refers to the ability of IoT systems (including hardware and software) to communicate, utilise, and collaborate on information. Due to the distributed nature of IoT systems and their heterogeneity, data interchange between different industrial sectors, strategic centres, and IoT systems is hard. As a result, IoT interoperability is difficult to achieve.

Privacy vulnerability The purpose of privacy is to ensure that IoT data is used appropriately and that no private user information is disclosed without the user's consent. Maintaining data privacy in IoT is difficult due to the complexity and decentralisation of IoT systems and the heterogeneity of IoT systems. Currently, IoT systems are being integrated with cloud computing to provide IoT with additional computational and storage capabilities. Uploading confidential IoT data to third-party cloud services might jeopardise privacy [21].

3.4 IoT on Blockchain

In addition to cryptocurrency, IoT is currently considered one of the fields where blockchain is needed. A decentralised peer-to-peer network architecture enables IoT device autonomy, and end-to-end communications do not have to go through a centralised server for performing automation services. To use blockchain technology to secure IoT devices, each device can have a unique address to send transactions. Thus, the IoT objects don't need to trust each other due to the use of the consensus algorithm, which allows nodes that connected to the blockchain to operate in a decentralised manner. Participants in blockchain networks can validate the integrity of the data delivered as well as the identity of the sender. Because no single organisation has control over the contents of a blockchain, IoT data and event logs recorded on the blockchain are immutable, ensuring accountability and traceability. In addition, the secure, tamper-proof storage in blockchains also enable the development of secure IoT applications. Integrating blockchain and IoT will add an extra security layer. This section showed how blockchain technology could improve IoT security and discussed several of the blockchain aspects that contribute to IoT security. Blockchain aspects that contribute to IoT security are as follows.

Distributed trust Blockchain distributes trust among different nodes in the blockchain network without the need for a trusted third party or intermediary to validate blockchain transactions. Rather than that, the validity and integrity of blockchain transactions are agreed upon through a consensus process. Additionally, because Blockchain is a distributed database with no single point of failure, each of its nodes receives a copy of all transaction records, with each transaction accompanied by a digital signature ensuring non-repudiation.

Confidentiality Blockchain technology offers an extremely high level of security for transactions recorded in the blockchain due to the encryption technology used to develop the blockchain. To help guarantee secrecy, the system will employ asymmetric encryption technology to ensure that all system access occurs only after authorisation.

Immutability/integrity Immutability is seen as one of the most significant benefits that the blockchain can provide for IoT systems. Once data is stored on the blockchain, altering or modifying it is exceedingly difficult. Thus, it ensures data integrity as it can demonstrate that the data has not been altered. As a result, the system will eliminate the need for a third party or other service providers to manage the verification process.

Availability Because each node receives a copy of all transaction records, Blockchain ensures a high level of availability for data stored on the Blockchain. Even if a node quits the network unintentionally, maliciously, or becomes inaccessible in some other way, the network as a whole will continue to function. As a result, a high level of availability is ensured.

Auditing Additionally, each Blockchain node is capable of proving any changes made to any of the Blockchain's linked blocks. This enables the tracking of data changes through the auditing and tracking of blockchain transactions. Additionally, this generates high-quality security intelligence regarding those transactions.

4 Blockchain Implementation in Different IoT Domain

Blockchain technology has the potential to be used in a wide variety of IoT applications. This section discussed the current state of blockchain deployment in various IoT domains.

4.1 Internet of Energy (IoE)

Internet of energy has shown considerable growth in the past few years, allowing the management of complex energy systems and making power grids smarter. Modern smart energy systems are also facing a range of cyber-attacks due to the automation and connection to the internet. Blockchain is currently proposed as a solution for many of the issues in IoE. For instance, the authors in [22] provided a systematic review of the challenges and opportunities for using blockchain in the IoE field.

They also classified blockchain applications for IoE based on different fields. Then discuss various case studies include peer-to-peer energy trading, e-mobility, electric charging, and the decentralised market. Similarly, the authors of [23] presented a blockchain-based protocol termed Directed Acyclic Graph-based V2G network (DV2G). The proposed approach utilises a tangle data structure to securely and scalable record network transactions.

4.2 Internet of Vehicles

Internet of vehicle have drawn more attention nowadays due to the increasing autonomy that has been given to the new generation vehicles. Opening new opportunities and potential to solve various traffic and road safety problems effectively. However, this has introduced new security challenges as it necessitates a secure data transmission and accurate data recording. To overcome some centralised issues and improve the security on the IoV, blockchain technology is introduced to provide a secure and decentralised vehicular environment. The work presented in Wang et al. [24] provided a comprehensive analysis of the applications of blockchain on the Internet of vehicles. The paper introduced IoVs and blockchain. Existing blockchain-based IoV surveys are also analysed. In order to better comprehend how the blockchain is implemented in IoVs, seven perspectives are investigated. Finally, the article discusses future blockchain-IoV research areas. Additionally, the authors of [25] presented an effective and secure distributed cloud architecture based on a blockchain instead of conventional cloud architecture in order to protect drivers' privacy while maintaining a low cost and on-demand sensing mechanism in the CRVANETs ecosystem. Similarly, the authors of [26] present another interesting work in which they proposed integrating blockchain technology with fog computing and cloud computing to achieve secure connectivity in automated vehicular networks. The proposed system can perform a variety of functions, including asset management, ownership sharing, cooperation, and collaboration among private Clouds.

4.3 Internet of Healthcare Things

Integrating IoT with healthcare allowed simultaneous reporting and real-time health condition monitoring via various smart medical devices, such as medical sensors and smartphone applications. However, the major concern about the electronic medical record is the security and privacy of personal health data. The authors in [27] analysed the potential of using the blockchain to secure medical health data that been recorded in the cloud. They also discussed the potential challenges of using blockchain to protect healthcare data. In [28] the authors proposed a smart health wallet that uses blockchain to give patients full access and control of their medical record. Another interesting work was presented in Liang et al. [29] where the authors combined blockchain with healthcare applications and introduced a solution for health data sharing. Their approach is to use blockchain to protect user privacy and to ensure the integrity of data collected and provided by a mobile application. To ensure the validity of health data recorded in the blockchain, the authors in [30] presented a secure attribute-based signature scheme that works with multiple authorities. Another approach presented in Liu et al. [31] which provided a privacy-preserving approach for medical health data exchange. Similarly, [32] makes use of blockchain to securely maintain medical record access policies.

4.4 Access Management in IoT

IoT networks can scale to connect millions of constrained devices, and one of the major challenges of having theses interconnected devices is the ability to manage them. The current centralised model has shown some limitations in managing this big number of devices. The authors of [33] proposed an IoT platform that based on hybrid and using blockchain called Log chain, and oneM2M s698tandard. Their prototype proves that the data collected from sensors recorded into private blockchain and can't be tempered. For the access management and control in the IoT, the authors in [34] introduced a new architecture for access control in IoT using private blockchain in Ethereum. In [35] the authors proposed a blockchain-based IoT update framework. This framework provides a secure verification method that uses blockchain to verify the firmware copy of the IoT devices distributed by the device manufacturer. In this case, the blockchain is used to ensure the integrity of the firmware. Another proposal is presented in Sun et al. [36] in which the authors proposed a blockchain-based IoT access control system that is secure and lightweight. The proposed system utilises a permissioned blockchain and attribute-based access control (ABAC) along with an identity-based signature (IBS). The proposed method involved the development of a policy decision point (PDP) selection algorithm capable of selecting multiple Internet of Things (IoT) devices (blockchain nodes) in real-time to achieve distributed policy decisions (off-chain).

4.5 Authentication and Identity Management

For identity management in the IoT system, the authors of [37] proposed a cryptographic membership authentication scheme for identity management which uses blockchain technology. The introduced system is used to links a digital identity to its real entity. Similarly, the authors of [38] proposed a new distributed A&A protocol based on blockchain technology for smart grid networks. The proposed approach secures power systems by leveraging the decentralisation and immutability properties of blockchain technology. The article proposes a blockchain technique for enabling both identity authentication and resource authorisation in smart grid systems. Another interesting work introduced in [39, 40], in which the authors present a novel authentication scheme for IoT-based architectures combined with cloud servers. To maximise efficiency, the proposed authentication scheme makes use of lightweight crypto modules such as the one-way hash function and exclusive-or operation. As a result, it is well-suited for resource constrained IoT devices.

4.6 Fog Computing

Fog computing stood out as an alternative to the cloud to extend its ability to serve geographically distributed IoT devices. However, this has introduced new security and privacy issues due to its characteristic which is defer from the traditional cloud in many different ways such as scalability, geographical distribution, and mobility [41]. For instance, the study presented in Bouachir et al. [42] provided the necessary knowledge of blockchain and fog computing to enhance the quality of service, data storage, computing, and security of cyber-physical systems. The article discussed blockchain technology for Cyber-Physical Systems in Fog Computing and used the smart industry as an example. In [43] the authors combined fog cloud service with blockchain to provide efficient resource management. The proposed model uses a fog cloud to manage the blockchain resources mainly the consensus process part. This help nodes with limited computational resources to interact with fog cloud and upload the computational part of the PoW to the fog cloud. In addition, the authors of [44] proposed a Bitcoin-based payment scheme for outsourcer fog users. This allows users to connect their resource-constrained devices to the fog cloud and pay for other fog nodes to do the computational tasks on their behalf.

4.7 Internet of Agriculture

Blockchain has also been used in the Internet of Agriculture and the food supply chain to provide reliable traceability. For instance, the study [45] thoroughly examines the value of integrating blockchain and IoT in the development of smart applications for precision agriculture. The study provided a review and discussion of the primary functions and advantages of the most widely used blockchain platforms for agriculture, including crops, livestock grazing, and food supply chain management. In addition, [46] presented AgriBlockIoT, a blockchain-based agriculture and food supply chain tracking solution, which allows IoT devices to record data on agriculture and food consumption and production. For the purpose of examining the various aspects of using blockchain and smart contracts on the Internet of Agriculture Things field, the work in [47] examined the various aspects of using blockchain and smart contracts in conjunction with integrating IoT devices in agriculture's pre-and post-harvesting segments. Additionally, the study proposed novel blockchain models that could be used to address significant challenges in IoT-based precision agricultural systems. The paper further discusses several of the security and privacy concerns and blockchain-related issues that impede the development of blockchain-based IoT systems for precision agriculture.

4.8 Software-Defined Networking

Software-defined for the IoT components implemented has been used to distribute IoT services to the IoT edge devices. In [48] the authors researched the idea of integrating the blockchain technology with the virtual IoT resources to provide software defined IoT components to edge hosts. Additionally, the authors of [49] proposed a secure SDN architecture for IoT called DistBlockNet. This architecture uses blockchain technology to provide verification and trust in IoT networks without the need for a third party. Another proposal [50] presented a blockchain-based security framework for vehicular IoT services, including real-time cloud-based video reports and trust management for vehicular messages. This article details the 5G-VANET model enabled by SDN, as well as the blockchain-based framework's scheduling procedures. Additionally, numerical simulation results demonstrate that malicious vehicular nodes or messages can be detected effectively while incurring minimal overhead and affecting network performance in large-scale scenarios.

4.9 Internet of Smart Cities

One definition of a smart city is a network of networked devices that capture, and process data supplied by people. However, assuring anonymity, completeness, and avoiding bottlenecks are still concerns [51]. The study presented in El Bekkali et al. [52] examined the Blockchain's suitability for ensuring the security of data transmitted and received by IoT network nodes. The proposed work demonstrated how to leverage blockchain technology to secure IoT systems and to create a secure decentralised architecture capable of providing a secure communication platform for smart cities. Another study [53] also presented a blockchain-based architecture, which proposed blockchain-based IoT solution for Air Quality Monitoring System (IB-AQMS). Similarly, the authors of [54] proposed a smart city architecture that can use blockchain technology to address cryptographic security issues. The proposed architecture preserves security issues such as confidentiality, integrity, and availability. Another proposal is presented in Li [55], which discuss blockchain applications in smart city, and introduced a blockchain solution for the security of IoT devices equipment upgrading and maintenance.

4.10 Intrusion Detection

The authors of [56] proposed a new model for event monitoring into the IoT blockchain network. The proposed framework represents an implementation of distributed pattern recognition which used for event monitoring into the IoT blockchain network. Another study is presented in Meng et al. [57], which provided an overview of integrating blockchain technology with IDS and discussed the ability of leveraging blockchain in securing intrusion detection processes. The paper has also discussed the challenges of such an approach. The work presented in Alexopoulos et al. [58] introduced blockchain-based IDS systems and discussed the uses of blockchain technology to improve trust between the collaborated IDS monitoring systems. Additionally, the authors of [59] proposed a consortium blockchain-based framework for detecting malware on mobile devices. Another interesting work was presented in Liang et al. [60], which involves designing and implementing an intrusion detection system that uses a hybrid method based on a multi-agent system, blockchain technology, and deep learning algorithms. Similarly, the authors of [61] introduced IDBIoT, an architecture for intrusion detection in an IoT environment. The proposed architecture is primarily concerned with network security by identifying intruders and preventing Goldfinger attacks using statistical significance.

4.11 Internet of Cloud

Cloud trust solutions have always relied on a centralised infrastructure, which results in significant administration overhead, network congestion, and even a single point of failure. In this context, blockchain offers novel solutions to Cloud of Things concerns such as decentralisation, data privacy, and network security. For instance, in [62] the authors proposed a blockchain-based data management model for cloud. The proposed module uses blockchain to provide transparency and trustworthy to data, which will enhance the security level in the cloud data center. Another study [63] provides a state-of-the-art review of blockchain integration with IoT (BCoT) in order to provide an overview of the BCoT from a variety of perspectives. The paper provided an in-depth examination of BCoT applications across a variety of use cases, including smart healthcare, smart cities, smart transportation, and smart industry. Additionally, the authors of [64] conduct a thorough examination of blockchain-based trust approaches in cloud computing systems. The paper identifies open challenges and suggests directions for future research in this field using a novel cloud-edge trust management framework and a cloud transaction model based on a double-blockchain structure. Another interesting work was presented in Sharma et al. [65], in which the authors proposed a blockchain-based distributed cloud module for IoT. This system uses a cloud fog network and a Software Defined Network (SDN) based on the blockchain. The proposed architecture provides high security and cost-effectiveness for IoT networks.

4.12 Blockchain for 5G

Global 5G wireless network deployment is approaching. 5G technologies connect heterogeneous devices and machines with considerable gains in service quality, network capacity, and system throughput. However, 5G systems still face security issues such as decentralisation, transparency, data interoperability, and network privacy. Traditional security measures may also be insufficient for 5G. Because 5G is often deployed in heterogeneous networks with many ubiquitous devices, secure and decentralised solutions are needed. On the other hand, blockchain and smart contracts may address some of the existing 5G network difficulties. For instance, the authors of [66] analysed the obstacles and issues surrounding the use of blockchain in a 5G-enabled IoT. They are then designed and evaluated a blockchain-based secure data management system for IoT communication entities. The proposed approach is immune to a range of potential attacks, which is crucial in the Internet of Drones (IoD) environments.

5   Blockchain Limitations

This section presents a thorough analysis of the existing challenges of integrating blockchain with IoT. Then discuss the currently proposed solutions.

5.1 Scalability

While public blockchains provide decentralisation, transparency, traceability, immutability and non-repudiation, these benefits come at the expense of a low transaction validation rate, high latency and storage space usage, thereby limiting public blockchains' scalability. This is due to the nature of the decentralised characteristics of the blockchain. As the blockchain network involves multiple connected nodes responsible for processing every transaction in the blockchain system, maintain a copy of the blockchain and all these nodes involved in the verification process. Thus, it takes a considerable time to reach a consensus and add a transaction to a block, making the blockchain un-scalable for applications that need faster access to the public ledger.

5.2 Throughput

The scalability of a blockchain system can be measured by comparing transaction throughput per second to the number of connected devices and concurrent workloads. Many blockchain systems are inefficient in terms of throughput. Substantial throughput restrictions are shown when comparing the throughput of Bitcoin and Ethereum transactions to that of other financial transaction systems such as VISA and PayPal. For example, Bitcoin can process approximately seven transactions per second on average [67], while Ethereum can manage up to 20 transactions per second [67]. On the other hand, VISA processes 1667 transactions per second and PayPal processes around 193 per second.

5.3 Energy Consumption

Since many blockchain systems, such as Bitcoin, are relayed on the Proof-of-Work (PoW) as the main consensus mechanism. The proof of work necessitates a tremendous amount of energy. According to Digiconomist statistics, the current Bitcoin's annual electricity consumption is about 73 TWh [68]. The estimated cost is above $16 million per day, and the estimated transactions' electricity consumption is 574 KWh. The electricity required for a single Bitcoin transaction is equivalent to the electricity required to power approximately 19.38 US households per day. Moreover, the carbon footprint per transaction is 272.44 KT of CO_2 [68].

5.4 Latency

One of the main challenges of integrating blockchain and IoT is the amount of time that it takes to create a block. In the Bitcoin blockchain, the confirmation can take up to ten minutes on average, while in Ethereum, it takes around 13 s in average to get confirmation. This is due to the mining delay; as every transaction takes place in the blockchain, it must wait on the queue with other transactions until a miner picks it up and generates a new block. As this is the nature of the work of the blockchain to achieve high security and prevent double spending, more time needs to be spent on the transaction validation process and block creation. Thus, it will be very hard to avoid. As a result, this might present an issue of integrating this technology with the IoT, especially for applications that require urgent access to the ledger or live update of the information collected from sensors into the blockchain [69].

5.5 Block Size and Bandwidth Related Issues

The size of blockchain systems, such as Bitcoin is presenting a continuous grows since their creation 2009. The current size of the Bitcoin blockchain is approximately 431.18 GB by the 28th of October 2021 [70]. While Ethereum blockchain has grown since its genesis block was first made available in 2015, and the current Ethereum database size is 345.17 GB, by 28th of October 2021 [71]. Moreover, the average Bitcoin block size is 1 MB. The time it takes to create 1 MB size Bitcoin block that holds around 500 transactions on average, is up to 10 min.

5.6   The Smart Contract Deployment Cost

One of the problems associated with building Distributed Apps on Ethereum blockchain is the cost associated with the creation of the smart contract. This can be one of the issues that effects building an IoT system that need to continuously update information to the public ledger. The cost of deploying and interacting with a smart contract to record temperature into Ethereum blockchain every second can be very high. The absolute deployment costs for a smart contract can be based on the size of the code of the smart contract, and the number of bytes that it assigns to the contract [72].

5.7   Usability

Nowadays, standardisation and diversity of infrastructure are presented a fundamental for the industry and financial adoption. However, there are many factors preventing blockchain from being widely adopted by the IoT industry. One of these issues is the lack of standardisation. The complexity is another issue, as there are many complex strings need it in order to secure your assets, such as seed, public and private keys. Additionally, there is a lack of regulatory certainty surrounding the underlying blockchain technology, which is a significant impediment to widespread adoption. For instance, complete transparency is at odds with confidentiality. Moreover, the immutability of data is a key feature of blockchain. Nonetheless, this feature may have a downside. The reason is that distributed ledger errors will be irreversible, and users will be unable to delete their personal data, which is inconsistent with the General Data Protection Regulation (GDPR) [73]. As a result, the issues mentioned may have a detrimental effect on the acceptance and investment in blockchain applications.

5.8   Currently Introduced Solutions

Scalability Different techniques have been proposed recently to improve the blockchain scalability issues. One such technique is blockchain storage optimisation, which solves the scalability issue by removing the old transaction records [74]. Another proposed technique for solving the scalability issue is Bitcoin-NG, which is main idea is to divide the conventional blocks into two parts, i.e. key block for leader election and micro blocks for storing different records [75]. Miners compete to become leaders, and leaders are responsible for generating micro blocks. This technique has also improved the longest chain approach in which only key blocks count and micro blocks do not carry any weight. This will help redesign blockchain and addresses the trade-off between block size and network security.

Energy Consumption A blockchain can be classified into three types: public, private, or hybrid. For instance, Bitcoin is public, and the mining consumes a significant amount of energy due to its reliance on a concept called proof of work [76]. As a result, the mining rigs must solve a deliberately complex cryptographic algorithm, which requires a significant amount of computational power. In addition, everyone on the network is competing to verify the block in order to earn bitcoins as a block reward. Private and hybrid networks do not require this type of mining as access is controlled through authorisation. Thus, nodes performing the verification can be "trusted," and the computations are relatively quick. These types of networks can be used to address the issue of energy consumption. Another way to solve the energy consumption problem is to reduce computations requirements produced from using Proof of Work. The studies [77] introduced moving from PoW to Proof of Stack (PoS).

Latency The type of consensus algorithm chosen for a specific blockchain causes the latency issue. The more complicated the consensus process, the longer it takes to process the transaction. The hashing algorithm used on blockchains also adds time to transaction times. Bitcoin-NG presents a byzantine fault tolerance protocol for improving the latency as compared to bitcoin [75]. Another approach was highlighted in [78], where the authors proposed the use of the Litecoin. Litecoin's blockchain is based on Script rather than the slower SHA-256 algorithm. As a result, transaction latency is reduced. Similarly, some other blockchain systems suggested moving from the chain of block data stricture to Directed Acyclic Graph (DAG) data stricture [79]. One example is the IOTA blockchain which using DAG data stricture in order to provide quick transactions.

Block size A lightweight client, such as in Ethereum [69] can alleviate block size and bandwidth difficulties by not storing a full copy of the blockchain. Then use API to interface with a blockchain node running on a proper computer device. Another solution is to use off-chain solutions [80] which are developed to increase the bandwidth of the network but also at the same moment increase the risk of data loss. In addition, while blockchain is a storage solution, it differs significantly from a database. Off-chain storage may help reduce storage costs. Data can be stored separately and linked to Blockchain using a pointer. Another work is presented in Hassanzadeh-Nazarabadi et al. [81], which proposed a blockchain architecture called LightChain, and it utilises a Distributed Hash Table (DHT) with participating peers. There are no permissions on the LightChain blockchain. Thus, all peers can easily access its addressable blocks and transactions.

Smart contract Smart contracts provide secure and reliable features for IoT which record and manage their interaction. One of the smart contract's main characteristics is that it has a way of enforcing or self-executing contractual clauses. This was technologically inviable until the advent of blockchain. Reduced cost, precision, speed, transparency and accuracy are the key advantages of smart contracts. Ethereum has been considered as the most popular smart contract blockchain platform [82]. However, the cost of deploying smart contracts and interacting with them has become a barrier to adoption. iOlite [83] introduced a smart contract solution to solve this problem by creating a Fast Adaptation Engine (FAE). This allows developers to write their own smart contracts. Through machine training, the FAE can understand complex programing language, which helps developers to create better smart contracts at no cost. In addition, smart contracts are vulnerable to different types of attacks as shown in Sect. 6, and hence require secure mechanisms for their efficient working. Verification mechanisms to guarantee their correct operations are required for their worldwide adoption by clients. Hackers are now exploiting vulnerabilities in smart contracts to steal enormous amounts of bitcoin. As a result, a high-security alert has been issued because several ERC20 token contracts are vulnerable to hacking. In order to solve the real-world security problems facing smart contracts written in Solidity, one impressive language called "Lity" [84] is being created for developing smart contracts across the CyberMiles blockchain, decentralised applications, and other customised blockchains. Lity aims to solve the performance and security issues facing Solidity. Loi and al. [85] Have also proposed a new solution called Oyente which is a tool that has the capability to track errors in any smart contracts. This can also be used to detect injection attacks as well as bugs in smart contracts. Moreover, Oyente achieves its task by analysing the smart contract's bytecode and follows the Ethereum Virtual Machine (EVM) execution model.

Security In [86] the authors proposed a quantitative framework that is used to analyse the performance and security provisions of the blockchain. It is a blockchain simulator and a security model that mimics its execution to evaluate basic security and performance. This model specifically focuses on the attacks of selfish and double-spending mining by taking into consideration the consensus protocol used and network parameters such as block propagation delays, block sizes, delays network, block rate and the mechanism of propagation of information, etc. in addition, Luu et al. [87] proposed a novel Smart Poll mining system, this is accomplished through the use of a smart contract, which can be used to ensure the transaction's security. Similarly, the authors of [88] proposed ByzCoin a novel consensus protocol based on Byzantine consensus. The proposed protocol has the ability to leverage scalable collective signing to submit transactions within a few seconds while also solve security issues. Two-factor authentication for a Bitcoin wallet has been proposed in Mann and Loebenberger [89] to solve the issue of authentication. In the two-factor approach, the wallet does not have to retain private keys but has a share of them. Thus, transaction can't be completed unless both devices have the share of the private key.

Data privacy Data privacy issues can be solved by encrypting the data. The authors of [90] proposed a technique for storing encrypted transactions and make use of compilers to translate the codes written by the programmers into cryptographically secure encrypted functions for hiding the information. Another technique named as Enigma [91] not only encrypts the data but also divide the encrypted data into small chunks and spread these chunks in the network in such a way that no node is able to access the data. We can also use other techniques for ensuring the privacy of data instead of data encryption. Off-chain solutions can be used in which data is stored outside the chain [92]. This kind of data is best to use when we must manage large amounts of data. In order to guarantee that only approved parties can access the data, different mechanisms are proposed for accessing the data.

Anonymity Many techniques are proposed to address the issue of anonymity. The most popular are Zerocash [93] and Zerocoin [94]. These approaches completely hide the sender and receiver's identities during a transaction, increasing privacy. Another approach proposed by Monero involves using a ring of signatures for making the records undetectable [95]. Similarly, Bitcoin Fog [96] increases privacy by using transaction mixing services. Such services divide the transactions into smaller transactions and make them obscure. Likewise, Coinjoin is another technique in which different users agree on joint payments and it could be presumed that transactions are from the same wallet [97]. These techniques try to improve bitcoin anonymity by adopting the concept of deregulating a bitcoin which is accused of promoting illicit operations.

6 The Future Research Directions

The integration of blockchain with IoT has opened up new directions for producing secure IoT applications and can be viewed as a starting point for generating new business models and decentralised IoT apps. Despite the tremendous advantages of blockchain technology, several critical challenges to solve before widespread adoption can occur, summarised as follows.

Blockchain-based IoT architecture One of the possible research directions can be the modification and designing of blockchain IoT architecture to meet the needs of IoT systems. We believe that the type of blockchain for different IoT applications needs to be optimised to meet the specific characteristics of IoT systems.

Privacy Blockchain technology is designed in such a way that all transactions are visible while also allowing for the identification of its participants. This is particularly problematic for public blockchains such as Bitcoin and Ethereum because the network ledger is publicly accessible, and all transactions are transparent and traceable. However, several protocols have been developed in the interim to provide an alternative to Bitcoin's pseudo-anonymity. However, additional contributions are required to establish new scalable blockchain systems for IoT.

Scalability The first blockchain generation has obvious scalability issues. For example, Bitcoin is limited to 7 transactions per second. Compared to Ethereum's 15 every second. Currently, faster chips, sidechains, and sharding are being investigated as answers to the scalability issue. However, this area still requires more contributions to developing new scalable blockchain systems for IoT.

Blockchain Consensus for IoT Currently, research are being carried out to overcome scalability limitations. These are being achieved by developing new consensus mechanisms, such as PoS, PoA, etc. Others recommended switching from blockchain to Directed Acyclic Graph (DAG) [79] like IOTA. However, the currently offered alternatives can boost throughput while compromising decentralisation and security. Due to the existing consensus mechanism's scalability limitations, further research should be taken to develop new consensus protocols that can reduce the time needed to reach consensus and create a new block while maintaining security.

Sharding The current approach to optimising consensus to fit with IoT systems and overcome the performance and scalability limitations in existing blockchain systems is sharding [98]. Sharding is a procedure that utilises a variety of different methodologies to classify blockchain nodes (shards). Nodes belonging to the same shard create a committee and collaborate to obtain consensus in parallel. However, as the number of nodes achieving the consensus is minimised, the probability of an adversary aborting the system becomes higher. Therefore, introducing new secure and scalable sharding methods remains an issue that needs further research.

7 Conclusion

This study provided an in-depth survey on the integration of blockchain and distributed ledger technologies with IoT, along with their architecture and main features, as well as technical working principles. The paper further reviewed the current research efforts that leverage the benefits of the blockchain in different IoT domains. Key challenges of integrating blockchain technology and IoT are discussed along with the currently proposed solution. The open research directions that can bring significant benefits and improvement to current blockchain systems to fit with current IoT applications were pointed. Finally, we concluded that integrating blockchain technology with IoT provides the potential of improving many IoT issues. However, many challenges need to be addressed before such integration. Based on the study's findings, we hope that this survey will serve as a reference and source of motivation for future research directions in this field.

References

1. Peña-López I (2005) The internet of things. In: ITU internet report (2005) Available online at. https://www.itu.int/osg/spu/publications/internetofthings/. Cited 20 Oct 2021
2. Nakamoto S (2008) Bitcoin: a peer-to-peer Electronic Cash System. Decentralized Bus Rev 21260
3. Buterin V (2013) Ethereum white paper. GitHub Repository 1:22–23

4. Ali MS, Vecchio M, Pincheira M, Dolui K, Antonelli F, Rehmani MH (2018) Applications of blockchains in the Internet of Things: a comprehensive survey. IEEE Commun Surv Tutor 21:1676–1717

5. Alam T (2021) A survey on blockchain and internet of things. Available at SSRN 3837964

6. Hui H, An X, Wang H, Ju W, Yang H, Gao H, Lin F (2019) Survey on blockchain for internet of things. J Internet Serv Inf Secur 9(2):1–30

7. Dai HN, Zheng Z, Zhang Y (2019) Blockchain for internet of things: a survey. IEEE Internet Things J 6(5):8076–8094

8. Wang X, Zha X, Ni W, Liu RP, Guo YJ, Niu X, Zheng K (2019) Survey on blockchain for Internet of Things. Comput Commun 136:10–29

9. Atlam HF, Azad MA, Alzahrani AG, Wills G (2020) A review of blockchain in internet of things and AI. In: Big Data Cognit Comput 4(4):28

10. Uddin MA, Stranieri A, Gondal I, Balasubramanian V (2021) A survey on the adoption of blockchain in IoT: challenges and solutions: blockchain: research and applications, 100006

11. AbuNaser M, Alkhatib AA (2019) Advanced survey of blockchain for the internet of things smart home. In: IEEE Jordan international joint conference on electrical engineering and information technology (JEEIT). IEEE, pp 58–62

12. Riabi I, Ayed HKB, Saidane LA (2019) A survey on blockchain based access control for internet of things. In: 15th international wireless communications and mobile computing conference (IWCMC). IEEE, pp 502–507

13. Zhu X, Badr Y (2018) A survey on blockchain-based identity management systems for the Internet of Things. In: 2018 IEEE international conference on internet of things (iThings) and IEEE green computing and communications (GreenCom) and IEEE cyber, physical and social computing (CPSCom) and IEEE smart data (SmartData). IEEE, pp 1568–1573

14. Chowdhury MJM, Ferdous MS, Biswas K, Chowdhury N, Muthukkumarasamy V (2020) A survey on blockchain-based platforms for IoT use-cases. Knowl Eng Rev 35

15. Hassanien AE, Elhoseny M, Ahmed SH, Singh AK (eds) (2019) Security in smart cities: models, applications, and challenges. Springer International Publishing

16. Pranto TH, Noman AA, Mahmud A, Haque AB (2021) Blockchain and smart contract for IoT enabled smart agriculture. PeerJ Comput Sci 7:e407

17. Ashton K (2009) That 'internet of things' thing. RFID J 22(7):97–114

18. Findstack [online]. Available at https://findstack.com/internet-of-things-statistics/

19. Patnaik R, Padhy N, Raju KS (2021) A systematic survey on IoT security issues, vulnerability and open challenges. Intelligent system design. Springer, Singapore, pp 723–730

20. Cheruvu S, Kumar A, Smith N, Wheeler DM (2020) IoT frameworks and complexity. Demystifying Internet Things Secur 23–148

21. Virat MS, Bindu SM, Aishwarya B, Dhanush BN, Kounte MR (2018) Security and privacy challenges in internet of things. In: 2018 2nd international conference on trends in electronics and informatics (ICOEI). IEEE, pp 454–460

22. Andoni M, Robu V, Flynn D, Abram S, Geach D, Jenkins D, McCallum P, Peacock A (2019) Blockchain technology in the energy sector: a systematic review of challenges and opportunities. Renew Sustain Energy Rev 100:143–174

23. Hassija V, Chamola V, Garg S, Krishna DNG, Kaddoum G, Jayakody DNK (2020) A blockchain-based framework for lightweight data sharing and energy trading in V2G network. In: IEEE Trans Veh Technol 69(6):5799–5812

24. Wang C, Cheng X, Li J, He Y (2021) Xiao K (2021) A survey: applications of blockchain in the Internet of Vehicles. EURASIP J Wirel Commun Netw 1:1–16

25. Nadeem S, Rizwan M, Ahmad F, Manzoor J (2019) Securing cognitive radio vehicular ad hoc network with fog node based distributed blockchain cloud architecture. Int J Adv Comput Sci Appl 10(1):288–295

26. Yin B, Mei L, Jiang Z, Wang K (2019) Joint cloud collaboration mechanism between vehicle clouds based on blockchain. In: 2019 IEEE international conference on service-oriented system engineering (SOSE). IEEE, pp 227–2275

27. Esposito C, De Santis A, Tortora G, Chang H, Choo KKR (2018) Blockchain: a panacea for healthcare cloud-based data security and privacy? IEEE Cloud Comput 5(1):31–37

28. Patel M (2017) Blockchain approach for smart health wallet. Int J Adv Res Comput Commun Eng 6(10):1–5

29. Liang X, Zhao J, Shetty S, Liu J, Li D (2017) Integrating blockchain for data sharing and collaboration in mobile healthcare applications. In: 2017 IEEE 28th annual international symposium on personal, indoor, and mobile radio communications (PIMRC). IEEE, pp 1–5

30. Guo R, Shi H, Zhao Q, Zheng D (2018) Secure attribute-based signature scheme with multiple authorities for blockchain in electronic health records systems. IEEE Access 6:11676–11686

31. Liu J, Li X, Ye L, Zhang H, Du X, Guizani M (2018) BPDS: a blockchain based privacy-preserving data sharing for electronic medical records. In: 2018 IEEE global communications conference (GLOBECOM). IEEE, pp 1–6

32. Tanwar S, Parekh K, Evans R (2020) Blockchain-based electronic healthcare record system for healthcare 4.0 applications. J Inf Secur Appl 50:102407

33. Lee C, Sung N, Nkenyereye L, Song J (2018) Blockchain enabled Internet-of-Things service platform for industrial domain. In: 2018 IEEE international conference on industrial internet (ICII). IEEE, pp 177–178

34. Novo O (2018) Blockchain meets IoT: An architecture for scalable access management in IoT. IEEE Internet Things J 5(2):1184–1195

35. Yohan A, Lo NW (2018) An over-the-blockchain firmware update framework for iot devices. In: 2018 IEEE conference on dependable and secure computing (DSC). IEEE, pp 1–8

36. Sun S, Du R, Chen S, Li W (2021) Blockchain-based iot access control system: towards security, lightweight, and cross-domain. IEEE Access 9:36868–36878

37. Lin C, He D, Huang X, Khan MK, Choo KKR (2018) A new transitively closed undirected graph authentication scheme for blockchain-based identity management systems. IEEE Access 6:28203–28212

38. Zhong Y, Zhou M, Li J, Chen J, Liu Y, Zhao Y, Hu M (2021) Distributed blockchain-based authentication and authorization protocol for smart grid. Wirel Commun Mobile Comput

39. Zhou L, Li X, Yeh KH, Su C, Chiu W (2019) Lightweight IoT-based authentication scheme in cloud computing circumstance. Futur Gener Comput Syst 91:244–251

40. Martínez-Peláez R, Toral-Cruz H, Parra-Michel JR, García V, Mena LJ, Félix VG, Ochoa-Brust A (2019) An enhanced lightweight IoT-based authentication scheme in cloud computing circumstances. Sensors 19(9):2098

41. Mukherjee M, Matam R, Shu L, Maglaras L, Ferrag MA, Choudhury N, Kumar V (2017) Security and privacy in fog computing: Challenges. IEEE Access 5:19293–19304

42. Bouachir O, Aloqaily M, Tseng L, Boukerche A (2020) Blockchain and fog computing for cyberphysical systems: The case of smart industry. Computer 53(9):36–45

43. Xiong Z, Feng S, Wang W, Niyato D, Wang P, Han Z (2018) Cloud/fog computing resource management and pricing for blockchain networks. IEEE Internet Things J 6(3):4585–4600

44. Huang H, Chen X, Wu Q, Huang X, Shen J (2018) Bitcoin-based fair payments for outsourcing computations of fog devices. Futur Gener Comput Syst 78:850–858

45. Torky M, Hassanein AE (2020) Integrating blockchain and the internet of things in precision agriculture: analysis, opportunities, and challenges. Comput Electron Agric (105476)

46. Caro MP, Ali MS, Vecchio M, Giaffreda R (2018) Blockchain-based traceability in agri-food supply chain management: a practical implementation. In: 2018 IoT vertical and topical summit on agriculture-Tuscany (IOT Tuscany). IEEE, pp 1–4

47. Pranto TH, Noman AA, Mahmud A, Haque AB (2021) Blockchain and smart contract for IoT enabled smart agriculture. Peer J Comput Sci 7:e407

48. Samaniego M, Deters R (2016) Using blockchain to push software-defined IoT components onto edge hosts. In: Proceedings of the international conference on big data and advanced wireless technologies, pp 1–9

49. Sharma PK, Singh S, Jeong YS, Park JH (2017) Distblocknet: a distributed blockchains-based secure sdn architecture for iot networks. IEEE Commun Mag 55(9):78–85

50. Xie L, Ding Y, Yang H, Wang X (2019) Blockchain-based secure and trustworthy Internet of Things in SDN-enabled 5G-VANETs. IEEE Access 7:56656–56666
51. Sookhak M, Tang H, He Y, Yu FR (2018) Security and privacy of smart cities: a survey, research issues and challenges. IEEE Commun Surv Tutor 21(2):1718–1743
52. El Bekkali A, Boulmalf M, Essaaidi M (2020) Towards blockchain-based architecture for smart cities cyber-security. In: 2020 international conference on electrical and information technologies (ICEIT). IEEE, pp 1–6
53. Benedict S, Rumaise P, Kaur J (2019) IoT blockchain solution for air quality monitoring in SmartCities. In: 2019 IEEE international conference on advanced networks and telecommunications systems (ANTS). IEEE, pp 1–6
54. Paul R, Baidya P, Sau S, Maity K, Maity S, Mandal SB (2018) IoT based secure smart city architecture using blockchain. In: 2018 2nd international conference on data science and business analytics (ICDSBA). IEEE, pp 215–220
55. Li S (2018) Application of blockchain technology in smart city infrastructure. In: 2018 IEEE international conference on smart internet of things (SmartIoT). IEEE, pp 276–2766
56. Hudaya A, Amin M, Ahmad NM, Kannan S (2018) Integrating distributed pattern recognition technique for event monitoring within the iot-blockchain network. In: 2018 international conference on intelligent and advanced system (ICIAS). IEEE, pp 1–6
57. Meng W, Tischhauser EW, Wang Q, Wang Y, Han J (2018) When intrusion detection meets blockchain technology: a review. IEEE Access 6:10179–10188
58. Alexopoulos N, Vasilomanolakis E, Ivánkó NR, Mühlhäuser M (2017) Towards blockchain-based collaborative intrusion detection systems. International conference on critical information infrastructures security. Springer, Cham, pp 107–118
59. Gu J, Sun B, Du X, Wang J, Zhuang Y, Wang Z (2018) Consortium blockchain-based malware detection in mobile devices. IEEE Access 6:12118–12128
60. Liang C, Shanmugam B, Azam S, Karim A, Islam A, Zamani M, Kavianpour S, Idris NB (2020) Intrusion detection system for the internet of things based on blockchain and multi-agent systems. Electronics 9(7):1120
61. Raja G, Ganapathisubramaniyan A, Anand G (2018) Intrusion detector for blockchain based IoT Networks. In: 2018 tenth international conference on advanced computing (ICoAC). IEEE, pp 328–332
62. Zhu L, Wu Y, Gai K, Choo KKR (2019) Controllable and trustworthy blockchain-based cloud data management. Futur Gener Comput Syst 91:527–535
63. Nguyen DC, Pathirana PN, Ding M, Seneviratne A (2020) Integration of blockchain and cloud of things: architecture, applications and challenges. IEEE Commun Surv Tutor 22(4):2521–2549
64. Li W, Wu J, Cao J, Chen N, Zhang Q, Buyya R (2021) Blockchain-based trust management in cloud computing systems: a taxonomy, review and future directions. J Cloud Comput 10(1):1–34
65. Sharma PK, Chen MY, Park JH (2017) A software defined fog node based distributed blockchain cloud architecture for IoT. Ieee Access 6:115–124
66. Bera B, Saha S, Das AK, Kumar N, Lorenz P, Alazab M (2020) Blockchain-envisioned secure data delivery and collection scheme for 5g-based iot-enabled internet of drones environment. IEEE Trans Veh Technol 69(8):9097–9111
67. Blockchain charts [online]. Available at https://www.blockchain.com/charts. Cited 20 Oct 2021
68. Bitcoin energy consumption index [online]. Available at https://digiconomist.net/bitcoin-energy-consumption. Cited 20 Oct 2021
69. Wood G (2014) Ethereum: a secure decentralised generalised transaction ledger 2017
70. Bitcoin (BTC) price stats and information [online]. Available at https://bitinfocharts.com/bitcoin/. Cited 20 Oct 2021
71. Ethereum (ETH) price stats and information [online]. Available at. https://bitinfocharts.com/ethereum/. Cited 20 Oct 2021
72. Sklaroff JM (2017) Smart contracts and the cost of inflexibility. U Pa L Rev 166:263

73. Suripeddi MKS (1964) Purandare P (2021) Blockchain and GDPR-a study on compatibility issues of the distributed ledger technology with GDPR data processing. J Phys Conf Ser 4:042005
74. Maroufi M, Abdolee R, Tazekand BM (2019) On the convergence of blockchain and internet of things (Iot) technologies. arXiv preprint arXiv:1904.01936
75. Eyal I, Gencer AE, Sirer EG, Van Renesse R (2016) Bitcoin-ng: A scalable blockchain protocol. In: *13th USENIX symposium on networked systems design and implementation (NSDI 16)*, pp 45–59
76. Zhang R, Chan WKV (2020) Evaluation of energy consumption in block-chains with proof of work and proof of stake. J Phys Conf Ser 1584(1):012023
77. King S, Nadal S (2012) Ppcoin: peer-to-peer crypto-currency with proof-of-stake. Self-published paper 19(1)
78. Bhosale J, Mavale S (2018) Volatility of select crypto-currencies: a comparison of Bitcoin. Ethereum and Litecoin, Annu Res J SCMS, p 6
79. Benčić FM, Žarko IP (2018) Distributed ledger technology: Blockchain compared to directed acyclic graph. In: 2018 IEEE 38th international conference on distributed computing systems (ICDCS). IEEE, pp 1569–1570
80. Poon J, Dryja T (2016) The bitcoin lightning network: Scalable off-chain instant payments
81. Hassanzadeh-Nazarabadi Y, Küpçü A, Özkasap Ö (2021) LightChain: scalable DHT-based blockchain. IEEE Trans Parallel Distrib Syst 32(10):2582–2593
82. Buterin V (2014) A next-generation smart contract and decentralized application platform. White Paper 3(37)
83. iOlite—smart contracts made easy [online] Available at https://medium.com/@iolite/iolite-smart-contracts-made-easy-2b94d982d41a. Cited 20 Oct 2021
84. Lity Documentation [online] Available at https://buildmedia.readthedocs.org/media/pdf/lity/latest/lity.pdf. Cited 20 Oct 2021
85. Luu L, Chu DH, Olickel H, Saxena P, Hobor A (2016) Making smart contracts smarter. In: Proceedings of the 2016 ACM SIGSAC conference on computer and communications security, pp 254–269
86. Gervais A, Karame GO, Wüst K, Glykantzis V, Ritzdorf H, Capkun S (2016) On the security and performance of proof of work blockchains. In: Proceedings of the 2016 ACM SIGSAC conference on computer and communications security, pp 3–16
87. Luu L, Velner Y, Teutsch J, Saxena P (2017) Smartpool: practical decentralized pooled mining. In: 26th USENIX security symposium (USENIX security 17), pp 1409–1426
88. Kogias EK, Jovanovic P, Gailly N, Khoffi I, Gasser L, Ford B (2016) Enhancing bitcoin security and performance with strong consistency via collective signing. In: 25th usenix security symposium (usenix security 16), pp 279–296
89. Mann C, Loebenberger D (2017) Two-factor authentication for the Bitcoin protocol. Int J Inf Secur 16(2):213–226
90. Kosba A, Miller A, Shi E, Wen Z, Papamanthou C (2016) Hawk: the blockchain model of cryptography and privacy-preserving smart contracts. In: 2016 IEEE symposium on security and privacy (SP). IEEE, pp 839–858
91. Zyskind G, Nathan O, Pentland A (2015) Enigma: decentralized computation platform with guaranteed privacy. arXiv preprint arXiv:1506.03471
92. Cheng R, Zhang F, Kos J, He W, Hynes N, Johnson N, Song D (2019) Ekiden: A platform for confidentiality-preserving, trustworthy, and performant smart contracts. In: 2019 IEEE European symposium on security and privacy (EuroS&P). IEE, pp 185–200
93. Sasson EB, Chiesa A, Garman C, Green M, Miers I, Tromer E, Virza M (2014) Zerocash: decentralized anonymous payments from bitcoin. In: 2014 IEEE symposium on security and privacy. IEEE, pp 459–474
94. Miers I, Garman C, Green M, Rubin AD (2013) Zerocoin: anonymous distributed e-cash from bitcoin. In: 2013 IEEE symposium on security and privacy. IEEE, pp 397–411
95. Wijaya DA, Liu J, Steinfeld R, Liu D (2018) Monero ring attack: recreating zero mixin transaction effect. In: 2018 17th IEEE international conference on trust, security and privacy in

computing and communications/12th IEEE international conference on big data science and engineering (TrustCom/BigDataSE). IEEE, pp 1196–1201
96. Moser M (2013) Anonymity of bitcoin transactions
97. Maurer FK, Neudecker T, Florian M (2017) Anonymous CoinJoin transactions with arbitrary values. In: 2017 IEEE Trustcom/BigDataSE/ICESS. IEEE, pp 522–529
98. Chow SS, Lai Z, Liu C, Lo E, Zhao Y (2018) Sharding blockchain. In 2018 IEEE international conference on internet of things (iThings) and IEEE green computing and communications (GreenCom) and IEEE cyber, physical and social computing (CPSCom) and IEEE smart data (SmartData). IEEE, p 1665

Quantum Bitcoin: The Intersection of Bitcoin, Quantum Computing and Blockchain

Yijie Zhu, Qiang Ni, Richard Jiang, Ahmed Bouridane, and Chang-Tsun Li

Abstract This survey briefly describes the background and development history of quantum bitcoin, introduces the relationship between blockchain and cryptocurrency, introduces in detail the structure of two quantum bitcoin systems and the advantages of quantum bitcoin compared with classical bitcoin, and discusses the advantages and possible challenges of the development of quantum bitcoin in the future.

1 Introduction: Background on Quantum Computing Meets Bitcoin

On November 1, 2008, a person named Satoshi Nakamoto published "Bitcoin: a peer-to-peer electronic cash system." [1]. The author whose real identity is still unknown described an e-cash trading system in this paper and shared its code open source. Since Nakamoto dug up the first bitcoin on January 3, 2009, bitcoin has been popular until today. Bitcoin is a P2P form of encrypted digital currency. As described in the paper, bitcoin does not have a separate entity like paper money or coins. As the title of the paper says, bitcoin is an e-cash system. The circulation of bitcoin depends on the network, which is more sufficient to complete transactions without a central institution. The issuance, payment and verification of bitcoin are completely independent and have no direct relationship with legal tender. The total amount of bitcoin is limited. It has no national boundaries and restrictions of the central bank. It is a currency that can circulate freely around the world in theory. Bitcoin system is a set of programs running around the world. Each distributed node maintains a

Y. Zhu (✉) · Q. Ni · R. Jiang
LIRA Center, Lancaster University, Lancaster LA1 4YW, UK
e-mail: y.zhu43@lancaster.ac.uk

A. Bouridane
Department of Computer Engineering, University of Sharjah, Sharjah, UAE

C.-T. Li
School of Info Technology Deakin, University Deakin, Geelong, VIC, Australia

© Springer Nature Switzerland AG 2022
R. Jiang et al. (eds.), *Big Data Privacy and Security in Smart Cities*,
Advanced Sciences and Technologies for Security Applications,
https://doi.org/10.1007/978-3-031-04424-3_12

certain size of ledger. Bitcoin uses the encryption system to protect all transactions in the ledger shared by all participants to form a so-called block, so it is called blockchain. A large number of Bitcoins are encrypted using asymmetric encryption methods. More specifically, the blockchain relies on ECC (Elliptic Curve Cryptography) for authentication. This encryption algorithm is almost impossible to crack for classical computers. However, quantum computing makes it possible to crack this encryption algorithm. The powerful computing power of quantum computer poses a great threat to the existing encryption algorithms. Although the quantum computer is still in its early stage and is still far from being truly practical, the potential threat of quantum computing to Bitcoin exists, and the practical quantum computer will make Bitcoin unsafe. In order to cope with this trend in the future, the concept of quantum Bitcoin is proposed [2]. A consequence of quantum theory is the No-Cloning Theorem which forbids arbitrary copying of quantum states [3]. No-Cloning Theorem refers to the process of exactly replicating any unknown quantum state in quantum mechanics, which is impossible, because the premise of replication is measurement, and measurement will generally change the state of the quantum.

As early as around 1970, Wiesner [4] and Broadbent and Schaffner [5] proposed a scheme to produce unforgeable quantum banknotes by using this theorem. Bennett and Brassard [6] published a pioneering paper in 1984, creating an independent field of quantum key distribution (QKD). QKD provides new possibilities for the realization of quantum money. However, due to the limitation of technology, the research of quantum money at that time stayed at the theoretical level and did not receive more attention. Until 2009, Mosca and Stebila proposed quantum coins [7]. Different from each banknote of quantum banknotes, quantum coins are exactly the same. This brings a new direction to quantum money. In 2009, Aaronson [8] proposed the first public key quantum currency system. In 2012, Aaronson and Christiano introduced a public key quantum currency based on hidden subspace [9]. Jorgenfors proposed an anonymous distributed currency: Quantum Bitcoin [2]. This is a Bitcoin like currency running on a quantum computer, using the No-Cloning Theorem. Compared with traditional Bitcoin, Quantum Bitcoin protocol has several advantages, including immediate local verification of transactions. Quantum Bitcoin only records newly minted money, which greatly reduces the occupied space and improves efficiency.

2 What's Blockchain and Cryptocurrency

Blockchain is an underlying technology derived from bitcoin [1]. Blockchain is a shared and unchangeable ledger used to record transactions and track assets in the network. Theoretically, almost anything that can be measured by value, including bitcoin, can be tracked and traded on the blockchain network. Blockchain has several key elements. The first is the distributed ledger, in which all users of bitcoin system can access the distributed ledger. Transactions made by users cannot be changed. The distributed ledger of bitcoin system records each transaction only once, avoiding the repeated work in the traditional transaction network. The second is the smart contract, which is a series of automatically executed rules stored on the blockchain to speed up the transaction [10]. A typical representative is Ethereum. There is also a record that cannot be tampered with. After the transaction is recorded in the distributed ledger, no transaction participant can change the transaction. If the record contains errors, a new transaction must be added, and then both transactions are visible to the user. The wrong transaction cannot be deleted. This technology is called blockchain. Whenever a transaction occurs, it will generate a data "block" to record the transaction and represent the flow of assets. Each block is connected to the blocks before and after it, and these blocks form a data chain with the change of asset location or ownership. Each additional block will strengthen the verification of the previous block, so as to improve the security of the whole blockchain. This enables the blockchain to prevent tampering and provides a key advantage that cannot be changed. As members of a private network, users can use blockchain to ensure that they receive accurate and timely data. Confidential blockchain records can only be shared with network members with special access authorization. All verified transactions are permanently recorded and cannot be tampered with. The shared distributed ledger eliminates time-consuming record reconciliation, and the smart contract can be executed automatically to speed up the progress (Fig. 1).

Cryptocurrency is a transaction medium that utilizes the standards of cryptography to guarantee transaction security and control the production of transaction units. Cryptocurrency is a sort of digital cash (or virtual money). The simplest cryptocurrency works by recording transactions into a database to compute how much cash every individual holds. In this sense, the framework isn't entirely different from the current activity method of banks. For instance, clients' assets spent online follow a rule: clients dispatch cash from a financial balance to one more record by deducting it from the number related with the client's record. This is just the data recorded in the database, and there will be no actual trade. Every cryptocurrency away at a similar premise since it is an enormous data information log, that is, a

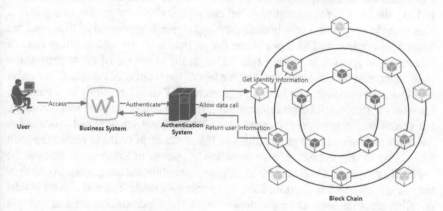

Fig. 1 A example of system with blockchain

transaction, used to decide the quantity of digital money ascribed to each address. The thing that matters is that cryptocurrency is absolutely digital, and there is no way to take out cryptocurrency as money or coins. Bitcoin is the first decentralized cryptocurrency. Sometimes people use Bitcoin to refer to all cryptocurrencies. The contrast between cryptocurrencies, for example, Bitcoin and conventional monetary models is the scattering of cryptocurrencies. This implies that when clients spend cryptocurrency, the endorsement of transactions comes not from a focal establishment, like a bank, yet from a point-to-point computer network. This is one of the most appealing and damaging parts of cryptocurrency. One view is that cryptocurrency can change the money related framework. Most cryptocurrencies additionally give security assurance, since everybody's privacy is taken cover behind the most exceptional cryptography, which implies that everybody's protection stays unaltered. Blockchain is the fundamental innovation of Bitcoin. After that, a large number of cryptocurrencies also adopt blockchain or similar technologies. At a time when the concept of Bitcoin is over extended, blockchain is often extended to too broad areas. However, for developers focusing on technology, blockchain generally refers to the combination of distributed ledger and decentralized network. The higher security and processing efficiency of blockchain in theory just accord with the characteristics of cryptocurrency.

3   A Brief Introduction to Quantum Computing

Traditional computers are made up of bits. These bits have two states: "0" and "1". Through the combination of these bits, the computer can achieve various complex functions. Quantum computers are different. Quantum computers are composed of qubits. In addition to "0" and "1", the state of qubit may also be between "0" and "1" at the same time, which is the so-called superposition state. $\frac{1}{\sqrt{2}}(|0\rangle + |1\rangle)$ is a representation of superposition state and the sum of the squares of the two states is 1. In the world of quantum physics, each qubit can be regarded as a particle. This particle is spinning continuously, the upper spin is expressed as $|0\rangle$, and the lower spin is expressed as $|1\rangle$. When the particle is in the superposition state of two states, it appears as "tilting spin". Due to the existence of the superposition state, compared with the two states of the bits of the traditional computer, the qubit of a quantum computer can have a large number of states, resulting in a quantum computer with a much larger information level than that of a traditional computer. For a traditional computer, n bits can encode 2^n states, and n qubits need to determine the respective occurrence probabilities of the 2^n states in order to obtain the joint probability of this system state so there are 2^{2^n} pieces of information that can be encoded by n qubits. This is a huge advantage over traditional computers. In addition, the process by which a quantum computer performs calculations and obtains results is called measurement. Through measurement, the superposition state of qubit is destroyed, qubit will collapse to $|0\rangle$ or $|1\rangle$ state. Perform multiple measurements on the same qubit. The collapsed state of this qubit is not related to the previous measurement, but is related to the aforementioned coefficients. To get the result, you need to perform multiple measurements to get the probability distribution of the two states. The "calculation" of a quantum computer is a physical phenomenon, which is different from the calculation of a traditional computer using registers, but a statistical probability. Using this feature, quantum computers can achieve faster processing speeds than traditional computers.

4 Grover's Algorithm, Shor's Algorithm and Impact on Bitcoin

RSA is one of the most generally utilized public key cryptosystems. It is generally a couple of RSA keys, one of which is a secret key, which is saved by the client; The other is a public key, which can be publicly unveiled and enlisted in the organization server [11]. The security of RSA calculation depends on the difficulty of large number factorization. As indicated by number theory, it is somewhat easy to track down two large prime numbers, yet it is incredibly hard to factorize their product. Hence, the product can be unveiled as an encryption key. To further develop the security strength, the RSA key is something like 500 bits in length, and 1024 bits are for the most part suggested. This makes the calculation of encryption exceptionally large. The specific description of RSA algorithm is as follows:

(1) Let $N = pq$ where p and q are prime numbers and $\varphi(N) = (p - 1)(q - 1)$;
(2) Let e be a number between 1 and $\varphi(n)$, $1 < e < \varphi(N)$, such that e is coprime to $\varphi(N)$;
(3) Let d be the inverse of e modulo $\varphi(N)$, $ed \equiv 1(\mathrm{mod}\varphi(N))$;
(4) (N, e) constitutes a public key, while (N, d) constitute a private key.

Many problems in reality can be summarized as search problems, such as brute force attacks on passwords. For traditional computers, the complexity required to solve this problem is O(N), while processing on a quantum computer using Grover's algorithm can achieve $O(\sqrt{N})$ complexity. The specific description of Grover's algorithm [22] is as follows:

(1) Start with the n-qubit state $|000...0>$.
(2) Apply the n-qubit Hadamard gate H to prepare the superposition state.
(3) Apply the Grover iterate a total of $\frac{\pi}{4}\frac{1}{\sqrt{N}}$ times.
(4) Measure the resulting state.

Shor's algorithm is a quantum algorithm with the end goal of Integer Decomposition [12]. It takes care of the accompanying issue: given an integer n, track down its prime factor. Shor's algorithm can be utilized to break the broadly utilized public key encryption technique, that is, RSA encryption algorithm. The premise of RSA algorithm is to accept an integer that can't be disintegrated efficiently. As of now, no classical algorithm is known to take care of this issue in polynomial time, however

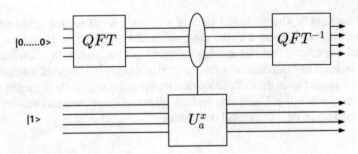

Fig. 2 Quantum circuit of Shor's algorithm

Shor's algorithm can solve this problem in polynomial time intricacy by utilizing quantum computer. Theoretically, an enormous enough quantum computer can break RSA. Shor's algorithm can solve such a problem: Given $N = pq$ for some unknown p and q such that $\gcd(p, q) = 1$, find p and q. The specific steps of Shor's algorithm are as follows:

(1)   Pick a random number, $1 < a < N$;
(2)   Compute $K = \gcd(a, N)$, the greatest common divisor of a and N;
(3)   If $K \neq 1$, then K is a nontrivial factor of n;
(4)   Otherwise, use the quantum period-finding subroutine to find the order r of a modulo N;
(5)   If r is odd, then go back to step 1;
(6)   If $a^{r/2} \equiv -1 (\mathrm{mod}\, N)$, then go back to step 1;
(7)   Otherwise, both $\gcd(a^{r/2} + 1, N)$ and $\gcd(a^{r/2} - 1, N)$ are nontrivial factors of N. In Shor's algorithm, quantum circuits are used to calculate the order (Fig. 2).

For cryptocurrencies including Bitcoin, because they use asymmetric encryption algorithms, such as RSA and Shor's algorithm, they can theoretically reverse engineer the private key, forge a digital signature, and then empty the user's Bitcoin wallet. Unlike Shor's algorithm, the threat of Grover's algorithm to Bitcoin lies more in cryptographic hashing. When cryptographic hashing is broken, the entire blockchain will be affected. In addition, using Grover's algorithm, miners can mine at several times faster than traditional methods. Even an attacker can generate a blockchain faster than the Bitcoin system to replace the real chain. Nevertheless, at this stage, the threat of quantum computing to the security of Bitcoin is still very limited. First of all, the computing power of quantum computers is quite limited. At present, the most advanced quantum computers such as D-Wave [13], Sycamore [14] and Jiuzhang [15] (Figs. 3, 4 and 5) are far from the number of quantum bits required to quickly crack asymmetric encryption algorithms, and the quantum algorithms that can be realized are even limited. The development of quantum computing is a gradual process and is gradually moving towards the target prospect, but this development is not instantaneous, which means that researchers have been in the process of establishing quantum security passwords, and the concept of quantum Bitcoin has also

Fig. 3 D-wave advantage quantum computer and QPU

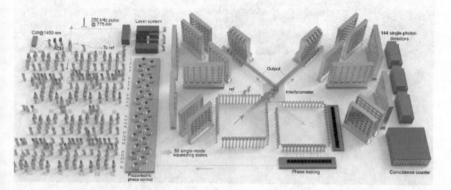

Fig. 4 JiuZhang-2 quantum computer

been proposed to deal with possible quantum cracking attacks in the future. The ability to resist quantum attacks can be improved by improving the Bitcoin protocol [16]. Another more comprehensive risk is to launch an attack from the elliptic curve signature used by Bitcoin. According to the most optimistic estimation, quantum computers may completely break the existing elliptic curve security as early as 2027 [17].

5 Possible: Quantum Bitcoin

The security of Bitcoin basically depends on cryptography based on the classical computer computing hardness assumption, and the latest development of quantum computer poses a serious threat to its security. Although there are still many difficulties to overcome if quantum computing is to be put into practice in a real sense, no one will doubt the potential danger. Compared with traditional Bitcoin systems, quantum-based systems try to solve the following problems: security, information

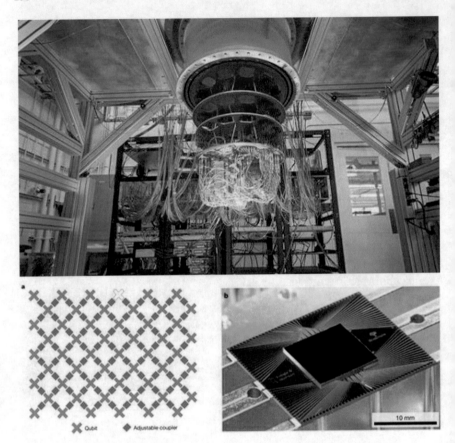

Fig. 5 Sycamore quantum computer and its QPU

transmission and improving the characteristics of Bitcoin itself. In order to counter the future quantum threat, many studies are being carried out.

Ikeda proposed a peer-to-peer quantum cash system called qBitcoin, which uses quantum methods to build Bitcoin like systems [18]. In Ikeda's idea, each individual qBitcoin is composed of a pair of classical states and quantum states. These classical bits and quantum bits (or qubits) are one-to-one correspondence. Different from the classical cryptocurrency in circulation now, the quantum information of each coin is unknown to everyone. After the quantum coin is distributed, the coin holder only knows the quantum state of the coin, but cannot obtain all the quantum information. For quantum systems, No-Cloning theorem [3] is the theoretical basis to prevent copying and forgery of quantum coin. Forgers cannot make coin by copying. As a cryptocurrency, qBitcoin needs to be able to allow everyone to verify the coin and serial number. When not all quantum information is open to the public, qBitcoin system uses quantum reporting to transmit quantum information. Assumptions$| \psi >$ It is the qBitcoin that the remitter wants to send to the payee. The remitter and the payee share an EPR pair [19]. First, the remitter pairs one of the EPR pairs and I ψ > Measure to obtain the quantum state, and then inform the payee of the result through the classical channel. The payee can use EPR to perform unitary operation to recover I ψ > Information about. The quantum state of the remitter is discarded and the coin is sent to the payee. In this way, the quantum information of qBitcoin cannot be saved by the remitter, and it is impossible to copy coin. In this process, quantum key distribution (QKD) protocol BB84 is applied [6]. In BB84 protocol, the sender arbitrarily sends two gatherings of single photons under non orthogonal basis vectors, and the collector haphazardly chooses the basis vectors for measurement. Ideally, when they utilize similar basis vectors, the two users will get a solid and steady key. The security of BB84 protocol is that two gatherings of non-orthogonal basis vectors are utilized for coding. Eavesdroppers cannot impeccably recognize these two groups of non-orthogonal quantum states without disturbance. As per the quantum No-Cloning theorem, the eavesdropper cannot impeccably clone the unknown quantum state, so the eavesdropper cannot listen in through the cloning activity. As per the quantum uncertainty principle, the eavesdropper's snooping activity will upset the first quantum state. The genuine correspondence groups can check whether there is an eavesdropper by contrasting data and one another, (for example, checking whether the mistake rate surpasses the hypothetical limit). Ikeda designed a trading system based on quantum chain for this system. The security of traditional Bitcoin digital signature is mostly based on the difficulty of solving mathematical problems. For example, RSA algorithm uses the huge amount of computation required to decompose a large number of factors to make cracking impossible. However, the application of quantum computer will pose a threat to such algorithms. Quantum digital signature may be applied to quantum systems in the future to deal with possible attacks in the future, such as the scheme proposed by Gottesman and Chuang [20] (Fig. 6).

Jorgenfors proposed an anonymous, distributed, secure Quantum Bitcoin based on quantum mechanism. Like Quantum Bitcoin, the core theory of quantum Bitcoin

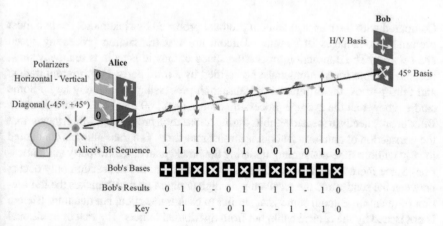

Fig. 6 QKD BB84 protocol

is the No-Cloning theorem. The protection of replication based on quantum system cryptocurrency is "built-in", which reduces the redundancy of the system compared with the classical cryptocurrency. Jorgenfors's scheme takes the quantum state as the monetary unit for settlement, and endows it with classical information for verification. Quantum Bitcoin is based on blockchain. Unlike some quantum currencies, Quantum Bitcoin does not require trust in the central bank. Classic Bitcoin requires every transaction to be recorded on the blockchain, which will take a lot of time, and the transaction of quantum Bitcoin can be completed immediately. Jorgenfors mentioned the main challenge of cryptocurrency based on distributed quantum system: untrusted miners. Quantum Bitcoin does not need a trusted central bank, which brings the advantages of speed and anonymity to Quantum Bitcoin. However, for individual miners, there is a phenomenon of "quantum double mining". Ideally, when a miner generates a new Quantum Bitcoin, it will generate a unique quantum state, but a malicious miner can reuse the private key to generate any number of identical Quantum Bitcoins after generating Quantum Bitcoins and adding them to the blockchain. For classic Bitcoin, its blockchain will record all transaction information. Therefore, once malicious miner hand over these copied Bitcoins to other owners, the system can immediately notice the abnormal conditions and handle them. However, for Quantum Bitcoin, without such transaction records, these copied Quantum Bitcoins will be easy to circulate. Considering the huge potential profits brought by this repeatedly generated Quantum Bitcoin, this threat needs to be paid attention to. In order to deal with this threat, coinage needs to be recorded in the Quantum Bitcoin system. Although there is additional overhead, the overhead of the protocol has been greatly reduced compared with the classical Bitcoin which stores all transaction records for dual consumption.

6 Discussion

Compared with the classical Bitcoin protocol proposed by Nakamoto, the two most prominent advantages of quantum Bitcoin are fast transaction processing speed and security. The transaction processing speed of classic Bitcoin is relatively slow, because Bitcoin transactions must be verified by a third party. For different traders and third parties, the difference of transaction processing is relatively large. Some studies show that the average processing time is about 60 min [21]. While quantum Bitcoin only needs to measure qubits during transaction processing, which means that the transaction of quantum Bitcoin is almost real-time. The transaction is conducted locally without time-consuming updating the blockchain and third-party authentication. Since there is no need for third-party verification, the transaction only occurs between the sender and the receiver, with high anonymity, which makes the transaction of quantum Bitcoin somewhat similar to cash transaction, but quantum Bitcoin is not issued by the central bank, but from distributed miners. The risk of traditional Bitcoin lies in the possibility of Bitcoin being copied, but for quantum Bitcoin, No-Cloning theorem eliminates this risk. Because the quantum Bitcoin system greatly simplifies the transaction process, there is no need to record the transaction time on the blockchain. In addition to the faster speed of generating the blockchain, it also occupies a small storage space, and the growth rate is easy to predict. In addition, since there is no need for the intervention of a third party, both parties do not need to pay additional transaction fees.

Quantum Bitcoin also has disadvantages. The most noteworthy problem is that quantum Bitcoin depends on quantum computers. Considering the use and cost of quantum computers, the number of quantum computers owned by early individual users may be relatively small, which means that most users need to use public quantum computers when trading quantum Bitcoin, such as the quantum computers of exchanges, But this is obviously contrary to the designer's assumption of local transactions of quantum Bitcoin. The existence of quantum double mining makes the whole trading system have some risks. Although the No-Cloning theorem brings great security advantages to the quantum Bitcoin system, once there are three or more participants in a single transaction, the transaction processing will become complex.

7 Conclusion

The contents introduced in this paper can be summarized as follows. With the development of quantum computing, the traditional Bitcoin trading system is in danger of being cracked. Therefore, a new Bitcoin system is needed to deal with this threat, and the quantum-based Bitcoin system is an option. Due to the existence of No-Cloning theorem of quantum system, quantum Bitcoin cannot be copied. Quantum Bitcoin does not require third-party verification, and its security and efficiency have been greatly improved. The local transaction of quantum Bitcoin can reach real-time transaction with high speed. At the same time, quantum still has disadvantages. For example, the transaction process of three or more participants will be more complex, which is a problem worthy of in-depth study in the future.

References

1. Nakamoto S (2008) Bitcoin: a peer-to-peer electronic cash system. https://Bitcoin.org/en/Bitcoin-paper
2. Jogenfors J (2019) Quantum Bitcoin: an anonymous, distributed, and secure currency secured by the no-cloning theorem of quantum mechanics. In 2019 ieee international conference on blockchain and cryptocurrency (ICBC), pp 245–252
3. Wootters WK, Zurek WH (1982) A single quantum cannot be cloned. Nature 299(5886):802–803
4. Wiesner S (1983) Conjugate coding. Sigact News 15(1):78–88
5. Broadbent A, Schaffner C (2016) Quantum cryptography beyond quantum key distribution. Des Codes Crypt 78(1):351–382

6. Bennett CH (1984) Quantum cryptography: public key distribution and coin tossing. Proceedings of IEEE International conference on computer system and signal processing, Bangalore, India, 175–179
7. Mosca M, Stebila D (2009) Quantum coins. ArXiv: Quantum Physics
8. Aaronson S (2009) Quantum copy-protection and quantum money. In 2009 24th annual ieee conference on computational complexity, 229–242
9. Aaronson S, Christiano P (2012) Quantum money from hidden subspaces. ACM symposium on theory of computing
10. Wood G (2013) Ethereum: a secure decentralised generalised transaction ledger
11. Rivest RL, Shamir A, Adleman L (1978) A method for obtaining digital signatures and public-key cryptosystems. Commun ACM 21(2):120–126
12. Shor PW (1999) Polynomial-time algorithms for prime factorization and discrete logarithms on a quantum computer. SIAM Rev 41(2):303–332
13. D-Wave Systems. https://www.dwavesys.com/. Availabled at 27 Oct 2021
14. Arute F, Arya K, Babbush R, Bacon D, Bardin JC, Barends R, Buell DA et al (2019) Quantum supremacy using a programmable superconducting processor. Nature 574(7779):505–510
15. Zhong H-S, Wang H, Deng Y-H, Chen M-C, Peng L-C, Luo Y-H, Hu Y et al (2020) Quantum computational advantage using photons. Science 370(6523):1460–1463
16. Sattath O (2020) On the insecurity of quantum Bitcoin mining. Int J Inf Secur 19(3):291–302
17. Aggarwal D, Brennen GK, Lee T, Santha M, Tomamichel M (2017) Quantum attacks on Bitcoin, and how to protect against them. Res Papers Econ
18. Ikeda K (2018) qBitcoin: a peer-to-peer quantum cash system. In Science and information conference, pp 763–771
19. Einstein A, Podolsky B, Rosen N (1935) Can quantum-mechanical description of physical reality be considered complete? Phys Rev 47(10):777–780
20. Chuang I, Gottesman D (2002) Quantum digital signatures. ArXiv: Quantum Physics
21. Karame GO, Androulaki E, Capkun S (2012) Double-spending fast payments in Bitcoin. In Proceedings of the 2012 ACM conference on Computer and communications security, pp 906–917
22. Grover LK (1996) A fast quantum mechanical algorithm for database search. In Proceedings of the twenty-eighth annual ACM symposium on Theory of computing, pp 212–219

Biometric Blockchain (BBC) Based e-Passports for Smart Border Control

Bing Xu, Qiang Ni, Richard Jiang, Ahmed Bouridane, and Chang-Tsun Li

Abstract The drastic shift from a centralized or federated system to decentralized system fundamentally changes the root of trust, which trust is derived from a system and the computing power attached to the system but not human any more. Blockchain, which currently is the most established decentralized peer-to-peer distributed system, sheds the light on a self-sovereign and password-free biometric identity management solution. By recording biometric identity template on the blockchain, a e-passport system is built upon BBC in W3C standardized digital identity format (JSON Linked Data). The tie between biometric identity template, one global unique person, and globally unique decentralized identifier are mutually and securely recorded as an immutable transaction on the blockchain, which make the tie cannot be repudiated. To preserve biometric data integrity and privacy, digital signature and biometric encryption is used.

1 Introduction

Border control is a very complex and time-consuming workflow, which mostly involves immigration and Customs control for people and goods entering a specific country. The identity credentials that are commonly used for verifying border crossing legitimacy include passports, visa, product packing list, and Customs declaration documents etc. Credentials are required to present at the border checking points and are used for both identity and legitimacy authentication and verification.

B. Xu (✉) · Q. Ni · R. Jiang
LIRA Center, Lancaster University, Lancaster LA1 4YW, UK
e-mail: b.xu3@lancaster.ac.uk

A. Bouridane
Computer Engineering, University of Sharjah, Sharjah, United Arab Emirates

C.-T. Li
School of Info Technology Deakin, University Deakin, Geelong, VIC, Australia

© Springer Nature Switzerland AG 2022
R. Jiang et al. (eds.), *Big Data Privacy and Security in Smart Cities*,
Advanced Sciences and Technologies for Security Applications,
https://doi.org/10.1007/978-3-031-04424-3_13

Fig. 1 UK border at Kent, 7000 lorry waits in a long queue to get across the border [1]

That task indeed is very time consuming and labor-intense especially when credentials are in tangible paper-based format. See Fig. 1 for a very long lorry queue jam waiting at UK Kent border. Plus, in 2003 when EU soft border zone suffers a sharp increase of illegal border crossing, an enhanced and more efficient border control solution is called. Thereafter, smart border control system, such as airport e-gates and Norway–Sweden 1600 km high-tech land border surveillant system, is deployed.

The central argument in smart border system is (1) how to verify a cargo/individual border crossing legitimacy as quick and accurate as possible. (2) the mechanism design of how to read and authenticate border crossing credentials efficiently and (3) how reliable those credentials are. So far, a major improvement has been made for passports, whose format already changed from only human readable paper-based passport to machine readable microchip-based passport. That is, Malaysia in 1998 was the first country which issues biometric e-passports, and in 2010, UK HM Passport Office issued an updated version 2.0 of e-passport, which holder's both facial image and soft identity that listed on the first page of the passport is stored in the chip.

The existing e-passport does not really improve border control workflow efficiency because the information in the chip is very limited (contains biometrics and soft identity only). Fast-pass e-gates are only available to the person who does not need a 'visa' to entering the country. However, people who needs visa to enter a country is required to attend human supported checkpoints to verify the border crossing legitimacy, which in that case costs the most of time and labor.

In the regard of security, conventional e-passport deploys asymmetric encryption such as digital signatures [2] and central authority certificates to protect the data integrity. However, as traditional e-passport chip is designed to be accessible and machine-readable, the data security becomes another major concern.

Considering above shortcomings, a remote soft border surveillant control solution that is privacy-aware biometric blockchain based framework at Edge is proposed. In this system, a decentralized digital biometric e-passport can be issued to both individual and vehicle, which aims to replace existing microchip-based e-passport with a W3C standardized digital-credential-based e-passport.

2 Literature Review

Blockchain, as an established decentralized and distributed network in current literature, draws exclusive attention since it is published in 2008 [3]. Blockchain has a very wide range of applications such as logistics [4], supply chain [5], healthcare [6, 7], education [8], and finance [9] etc. Distributed system like blockchain has very distinguishing advantages in heterogeneous data resources collection over both data object and a service of presenting a procedure interface to the user due to its intrinsic resources management mechanism [10]. By blockchain system intrinsic architecture, the system has many favorable build-in features, such as decentralization, distributed, disintermediation, immutable, irreversible, and trust-worthy [3]. Most important, once a transaction is validated and broadcasted in the blockchain, the transaction records will be permanently and immutably recorded on the blockchain [11]. Inspired by that, an immutable and persistent tie can be built up between a biometric identity template and one unique identifier (a.k.a. blockchain header or TXID) if the biometric template can be recorded as a blockchain transaction.

The main challenge in BBC based identity management (IdM) system is two folds. First, biometric data privacy and security. Biometrics is not private information, such as gaits, facial image, and voice. That biometric data is publicly available and collectable even without the notice of the biometric identity subject. Plus, biometric data has a significant amount of inter- and intra- personal variance, which makes direct encryption impossible. Most important, blockchain as a distributed system, all nodes on the network are fully informed about every details of all transactions [12]. Therefore, extra precaution has to be given to protect biometric confidentiality and privacy. Second, cost and performance. Biometric identity recognition based on distinctiveness and persistency of the biometric feature, which requires a good amount of computation power. BBC is expensive as every calculation in the network is charged a fee. Therefore, a good practice of BBC based IdM is normally a trade-off between cost, security, and privacy.

Countermeasures of above challenges include biometric encryption [13], biometric as private key [14], cancellable biometrics [15], off-blockchain storage in interplanetary file system (IPFS) [16], smart contract enabled biometric authentication [17], and zero knowledge proof [9, 18] etc. Specifically, [14] suggests to use biometric encryption to extract a fuzzy biometric key that can be used to sign a signature in the blockchain system, but the drawbacks in it is the public key file size is ten times larger than default classic blockchain system and the signature generation is around 16 times slower. [19] suggests a practical biometric signature scheme in blockchain; however, the cost of collecting and authenticating user signature becomes too high to be adopted at a general public scale.

There are some established decentralized identity applications in current market, such as UPort [20], ShoCard [21], and Sovrin [22]. They all build upon decentralized network and generate user fully self-sovereign identities [23]. Both UPort and Sovrin are open sourced, but ShoCard is not. It targets to improve airline customer travel experiences, which is very similar with our proposal. Specifically, ShoCard [21] is a token-based and decentralized travel identity wallet with built in facial recognition mechanism. It aims to enhance air travel customer experience and improve efficiency at traveler checkpoints. ShoCard in general is a public key infrastructure [24] where its server takes the responsibility to transfer messages between ShoCard user and airport agencies (public key certifier). Major drawbacks in ShoCard is facial image still needs to transmit between certifier and user. Even though the data is encrypted in envelope, there is still a risk for man-in-the-middle attack. Plus, if certifier stop issue certificates to user, the corresponding identity loses credibility completely.

3 Preliminaries

3.1 Blockchain

Blockchain applies cryptography to distributed network and build a timestamped chain of hashed blocks. All transactions [11] in blockchain are irreversible, immutable, transparent to the public and timestamped in Unix epoch time, which makes the system itself can be intrinsically trusted. Even though irreversible is not favorable when disputes happen among users, it dramatically reduces transaction cost as none needs to pay for intermediary fee. Blockchain system can has as many users as possible. Users, who help the blockchain network maintain its consensus rule, are called miners. Miners validate blocks and prove the validation of a transaction. Blockchain as its name which is consist of blocks. Each block contains [3]:

(1) Block header. Block header is a substructure of each block, which has previous hash, nonce, time, nBits, and Merkle root hash inside. Block header is also an alphanumeric unique identifier of each specific block, in that case block header is also called block hash and is generated by formula (1). BH is the block header, C is the concatenated all raw information in current block header, and SHA256 is the encryption algorithm.

$$BH_n = SHA256(BH_{n-1}|C_n) \tag{1}$$

(2) Merkle root hash. Hash in internal byte order of all transaction identity (TXID) in the same block. That is, $H(H(TXID0|TXID1)\ |H(TXID2|TXID3)\ ...)$, and H is the hash function such as SHA256. Only a maximum of two TXIDs can be concatenated into 64 raw bytes at one time, concatenation will be repeat constantly until one final root hash is generate.

(3) Previous hash. Previous hash is exactly the same with the block hash of the previous block, which maintains the chain rule in blockchain, and is hashed in internal byte order.

(4) Target nBits. A target 256-bit unsigned integer threshold which is a standard for block header validation. That is, a valid block header must be not larger than target nBits.

(5) Nonce. Miner valid a block by finding a proper block header through brute force trial and error. Nonce is the minimum number that a miner has to try before getting a proper block header and make it valid.

(6) Time. A Unix epoch time when a miner begins to hash a block header, which must be less than or equal to the median time of previous 11 blocks.

3.2 Verifiable Credential and Decentralized Identifier

Credentials are indispensable part of our life. Like for instance, driving license proves a person is capable of driving vehicle, university degree certificate proves a person holds a level of education, and a passport declares a person's nationality etc. A physical tangible credential in most cases contains the information of (1) issue authority. (2) mechanism of identifying the subject of the credential(identifier, name, or photo etc.) (3) the type of the credential (bank card, driving license, passport etc.) (4) attributes or features being asserted by the issuing authority about the subject (nationality, date of birth, vehicle entitled to drive etc.) (5) constrains on the credential (expire date, or terms of use) (6) Evidence of the origin of the credential(anti-forgery label etc.) [25].

The format of identity credentials is changing with time and computing technology. When conventional physical tangible or scanned copy of a credential is no longer capable for application over the internet, fully digitalized and verifiable digital identity credential is standardized and recommended by W3C in 2016 [25]. In accordance with W3C 'Verifiable Credentials Data Model v1.1' recommendation, fully digitalized verifiable credentials can express exactly the same information as a physical tangible credential, and at the same time be even more tamper-evident and more trustworthy than the conventional credential counterparts by deploying additional technology like digital signatures [25].

In a similar fashion, identifiers are syntax that we use to identify a subject in a variety of contexts, like communication identifiers as in mobile phone number and email address, identity identifiers as in passport number, driving license, or insurance number etc. These identifiers are usually unique within its predefined scope and have a wide range of variety. Identifier indeed does not have to embed the identity of the referent, but only accomplish the task of identification an entity regardless of method [26]. In conventional scenario, these identifiers are either centralized or federalized. They do not belong to ourselves but given and controlled by those identifier issuers. That is, we are identity users but not owners. For example, UK passports are solely issued by Her Majesty's passport office. UK citizens can use passport to prove who they claim to be, but the HM passport office has the full control over the passport and can revoke and modify the passport at any time.

As a comparison, decentralized identifier (DID) is a brand-new form of identifier that is defined by World Wide Web Consortium (W3C) organization in 2000 [26]. Every DID has three components: (1) a controller who generates and owns the decentralized identifier; (2) a subject which is what the DID identifies, and (3) a corresponding DID document which contains the mechanism enable DID controller to prove their ownership to-wards a specific DID [27]. Due to the fact that a digital identity consists of a unique identifier and its corresponding credential, decentralized identifier makes verifiable, persistent, and decentralized digital identity possible so that entities like individuals or organizations can generate their own DID in a trustable system.

DID belongs to a subclass of uniform resource identifier (URI), but is compulsorily registered on decentralized network or distributed ledger technology-based registry. Sporny et al. [26] has W3C official detailed DID generic syntax rule, and [27] lists the potential use-case of DID. The benefits of deploying W3C DID standard to issue identifier are three folds. First, improve interoperability. Second, easier implementation. Third, established anti-threats model. Most important, decentralized identifier documents are usually stored off-distributed network; therefore, it is more secure compared with smart contract-supported decentralized identity system.

4   The Proposed e-password System

This proposal of remote smart soft border control regime is based on three major components: (1) individual and vehicle e-passport; (2) real time soft border surveillance system for live data collection and reconciliation; and (3) border crossing events auto recording (BBC Virtual Stamping on e-passport). The novelty of this proposal are two folds. First, the proposed e-passport is suitable to replace existing microchip-based e-passport, which our new e-passport is totally intangible, persistent, and globally unique. Plus, secured by blockchain and biometric fuzzy extractor, proposed e-passport is revocable, auditable, and all events history is immutable. Second, the e-passport is applicable to both individual and vehicle, and both identity and border crossing legitimacy can be remotely authenticated without the necessary to go for a checkpoint.

4.1   Main Entities

(1)   *Border Agency (BA)*

Border agency is the only trusted authority in the system, who initializes the system, collects and store user both biometric data (fingerprint and facial image) and soft identity data (name, date of birth, and gender), maintains user e-passport, audits border crossing records, and revoke user e-passport whenever it is necessary. Most important, BA collects trustable information from motor vehicle department, and Her Majesty's Revenue and Customs for further border pass-through reconciliations, and linked this verifiable information as conditional attributes to the verify user border crossing legitimacy. BA maintains blockchain.

(2) *Motor Vehicle Department (MVD)*

The motor vehicle department will collect driver soft identity information, driving plate number and vehicle color, insurance information etc. MVD then sends this information to BA so that BA can link vehicle information with its corresponding driver's e-passport. Vehicle owner has the flexibility to choose if their vehicle is to link with their e-passport or not. However, in the case of a vehicle is going to be used for border-crossing, driver has to include vehicle information in their e-passport.

(3) *Her Majesty's Revenue and Customs (HMRC)*

HMRC gives legitimate permissions to vehicles to pass-through the border with declared loading products. HMRC takes the responsibility to send the declared product details (lists of product, weights, driver's details and transportation vehicle details) to BA. BA will reconciliate these declared claims later when relevant vehicle pass-through the border under surveillant system (poetical camera, actuators, and sensors' readings).

(4) *Blockchain*

Two separate blockchain will be generated in the proposed system. One blockchain is permissioned Hyperledger, which BA takes full control over it and maintains user decentralized e-passport. The other block-chain is built on top of Ethereum, which auto records all border crossing events on the blockchain as a transaction.

(5) *Surveillant System*

Surveillant system consists of optical cameras, weight reading sensors, and x-ray scanners, which aims to collect vehicle driving plate number, vehicle color, real time weight, and visual check vehicle loaded products. All real time readings will be sent to BA for further reconciliations.

4.2 Biometric Authentication

Biometric identity authentication has four established procedures: (1) Raw biometric data collection. Biometric data needs to be collected first through sensor, camera, or video devices as candidate biometric vector (CBV). Biometrics data contains a lot of intra-and inter-personal variance due to data collection environment, devices, and positions etc. (2) Identity feature extraction. Biometric features will be extracted from CBV through feature extraction algorithms. Popular fingerprint feature extraction algorithms are stacked autoencoder, convolutional natural network, and restricted Boltzmann machines. (3) Template generation. Extracted features in most cases will be encrypted and then stored in database as initial biometric vector (IBV) which serve as the biometric template. Classical biometric template protection methods include biometric encryption [28] and cancellable biometrics [15]. Due to the fact that biometric sample varies a lot at each time of collection, a major technology challenge of biometric encryption is how to regenerate the exactly same key from different input. To tackle the nature variation in biometric data, fuzzy algorithms, like Fuzzy Commitment [29] and Fuzzy Vault [30], are generally applied to biometric encryption. (4) Matching. Whenever a biometric identity authentication request is raised, a fresher CBV will be collected from identity claimer to match with IBV in biometric template database. In most cases, biometric identity verification is one-to-one match, and the result of that depends on a cost function to be minimized. That is, $X = (x_1, x_2, \cdots, x_n)^T$ is assumed as the extracted biometric template feature vector, and $X' = (x'_1, x'_2, \cdots, x'_n)^T$ is a freshly extracted biometric feature vector. The matching metric rule is to calculate the distance by a formula of $dist(X, X') = \left(\sum_{i=1}^n (x_i - x'_i)^2\right)^{1/2}$.

To extract a cryptographic fix key string from biometric raw data to function as private key in blockchain network, fuzzy extractor can be used. Fuzzy extractor fundamentally is a probability function model. In accordance with [31], the supporting knowledge behind the fuzzy extractor is secure sketch and strong extractor, which in together makes noisy input exact reconstruction possible. That is, suppose M is a strong randomness extractor, and its function is expressed as Ext: $M \rightarrow \{0, 1\}^l$ with r randomness. $(m, l, \in)$ is called a strong extractor if for all m-sources W on $M(Ext(W, I), I) =\in (U_l, U_r)$, and $I = U_r$ is independent of W. As for secure sketch, suppose any noisy input raw data w and its corresponding sketch s, given s and a value w'. If w' is close enough to w, w can be exactly recovered. However, one strong requirement in secure sketch is s can not expose any information about w at all. For a mathematic expression, secure sketch is:

(1) Suppose the noisy input is w, and the sketching procedure is S. S operates on w, and generate a helper string $s \in \{0, 1\}^*$.

(2) To exactly recovery input w, element w' is collected. If $dis(w, w') \leq t$, then $Rec(w', s(w)) = w$ and Rec is a recovery correctness function.

Instead of exactly recover the noisy input, fuzzy extractor [32] is to precisely recovery an input data generated data string. For a detailed explanation of fuzzy extractor, for any biometric raw data set B, sets a uniform distributed random numeric string R (PIN/parameters/secrets). As long as fresher collected biometric raw data set B' is close enough to B, R can be exactly recovered from B'. To put in more detail, a fuzzy extractor $(m, l, t, \in)$:

(1) Generate (Gen). Suppose $w \in M$, which M is a metric space with distance function $'dis'$. Gen outputs an extracted string $R \in \{0, 1\}^l$ and a helper string $P \in \{0, 1\}^\varepsilon$.

(2) Reproduce (Rep). If $dis(w, w')$ $\leq$ t $and(R, P)$ $\leftarrow$
$Gen(w), then\ Rep(w', P) = R.$

The output of above is R, which is uniform random sequences of bits and commonly is used as cryptographic secret keys.

4.3 System Modelling

The whole system generates two BBCs. One BBC is for generating e-passports for both individual and vehicle, and the other BBC records all border crossing events as a virtual stamp stamps on the e-passport. Essentially, this system is to copy original hard border checkpoints workflow but only put the detailed procedures into remote online process as much as possible. Our new e-passport can be easily and non-repudiate linked with other necessary and compulsory border crossing legitimacy credentials, which aims to maximize the functionality and interoperability of e-passport. Virtual stamps are only granted after all authentication is completed, including legitimacy verification, biometric identity authentication, and remote surveillance monitors and reconciliation. See Fig. 2 for a demonstration of our proposed system workflow.

BA takes the responsibility for maintaining both BBCs and monitoring the surveillant system. In accordance with data flow, system model can be descripted into three main procedures:

(1) Raw data collection. Raw data comes from three channels. First, e-passport application through BBC VID generation website. In this channel, all e-passport applicants are required to visit BA site to enable BA collect their biometric data. Second, surveillant system Edge IoT, including CCTV monitors for vehicle driving plate number and vehicle color, road actuators for vehicle weights, cargo X-ray scans etc. Third, mobile Edge devices like mobile phones. Individual pushing border crossing legitimacy verification request via mobile Edge server. In this channel, e-passport holder's fresh biometric data (facial image and fingerprint) will be recollected, linked border-crossing legitimacy documents (document number), real time GPS location etc. will be collected.

(2) E-passport generation and linked credentials. BA stores e-passport holder's biometric raw data in its own database, and then uses fuzzy extractor to generate a fixed PIN. The PIN will be concatenated with hashed soft identity, and then the concatenated result will be hashed again. The final result is a 256-bit long string and will be pushed into blockchain as a transaction with BA signed signature attached to transaction. The result of this operation is a TXID, which is also the decentralized identifier of the e-passport. TXID can be dereferenced into a BA signature and the 256-bit long hashed result of fuzzy extracted PIN and soft identity, which is the specific content of the e-passport. Lined credentials are represented as a credential number which is given by other main entities (HMRC and DVLA).

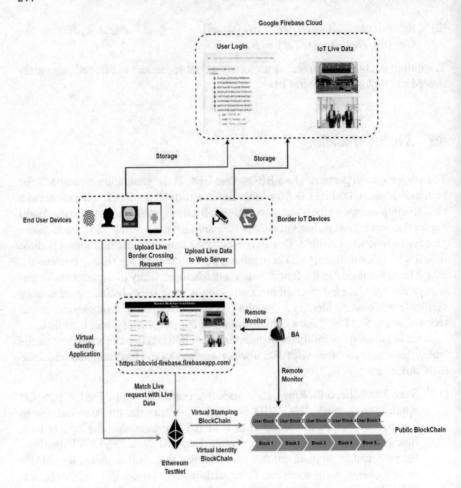

Fig. 2 Diagram of proposed remote soft border control workflow

(3)   Data reconciliations and virtual stamping BBC generation. The virtual
      stamping BBC is generated by BA but is requested by e-passport holder.
      E-passport holder sends border crossing request to BA by filling a mobile
      application-based form and submitting all necessary data for authentication.
      BA reconcile and authenticates all data at a web-based channel and leave a
      stamp (transaction history) by pushing the authentication results to the BBC.

5 Implementations and Performance

(1) E-passport application and registration

Individual and vehicle e-passport applicant has to visit a website to make an application for BA to collect all necessary information such as applicant name, date of birth, place of birth, visa information, and existing tangible paper-based passport information etc. After submission, one application number will be generated.

After registration, individual applicant has to visit BA to submit biometric identity information in person on site. After data collection, applicant is requests to set up a PIN that will be linked with the biometric data. BA officer needs to visit e-passport generator website to upload applicant information including biometrics and pre-collected soft identity to web server, and sign a signature by BA's own private key. In accordance with W3C digital identity standard format, e-passport is generated as in a decentralized identity in JSON Linked Data (JSON-LD) [33]. Specifically, the e-passport generator will be encrypted all information first and then push it to blockchain network. The blockchain network will generate a TXID and a signature file attached to the transaction. The TXID specifically is the e-passport identifier in the system, and the transaction context is the context of the e-passport.

(2) Biometric identity authentication and Surveillant Information Reconciliation

There is a mobile application developed as 'BBC VID' (Biometric Blockchain Virtual identity) for user to upload border crossing legitimacy documents and make identity authentication request to BA server. User needs to fill all relevant information like e-passport number, loading vehicle information (if any), product list number (if any) etc. so that BA can verify the border crossing legitimacy. There are four components in the mobile application. 1st, legitimacy information collection. Custom documentation file number like product list and Custom declaration number; Vehicle e-passport number and color, user e-passport and visa, which all needs to be collected for border crossing legitimacy verification. 2nd, real time GPS location collection. This is to know about the position of the vehicle and applicant at the moment of verification. It should be able to be reconciliated with BA on site surveillant CCTV system records. 3rd, biometric identity authentication. A real time face image and fingerprint data collection function is built in the application. Once face image and fingerprint identity are both authenticated. The mobile wallets [34] will generate a one-ff private key for the user. Forth, after all above three procedures are completed, user can use 'generate time stamps' button to push all information to BA server and sign this request as a transaction on blockchain and a signature is attached to it. BA server will generate a virtual stamping record and return to user mobile application wallets once border crossing legitimacy is verified.

6   Conclusion

This proposed system is built upon Hyperledger Indy private blockchain. Because the underlying network is only a test net, we use both prove of work and prove of authority as the consensus rule. Limited by the availability of biometric sample and mobile application users, we found it is very difficult to measure the performance in a large scale. The system can be improved in the regard of building a more profound connection between all involved official organizations like a consortium system, so that user border crossing legitimacy can be verified more efficiently. See below Fig. 3 for a demonstration of e-passport application and e-passport generation website.

The advantages of this proposed system are all transactions are recorded in blockchain. Both individual and vehicle e-passports, and their border crossing events history, blockchain makes these history records immutable. Plus, biometric identity is intrinsic to one and only person at a global scale, the persistent and non-repudiate relationship between the e-passport and the person can be strongly built up. See below Fig. 4 for a demonstration of the e-passport mobile application.

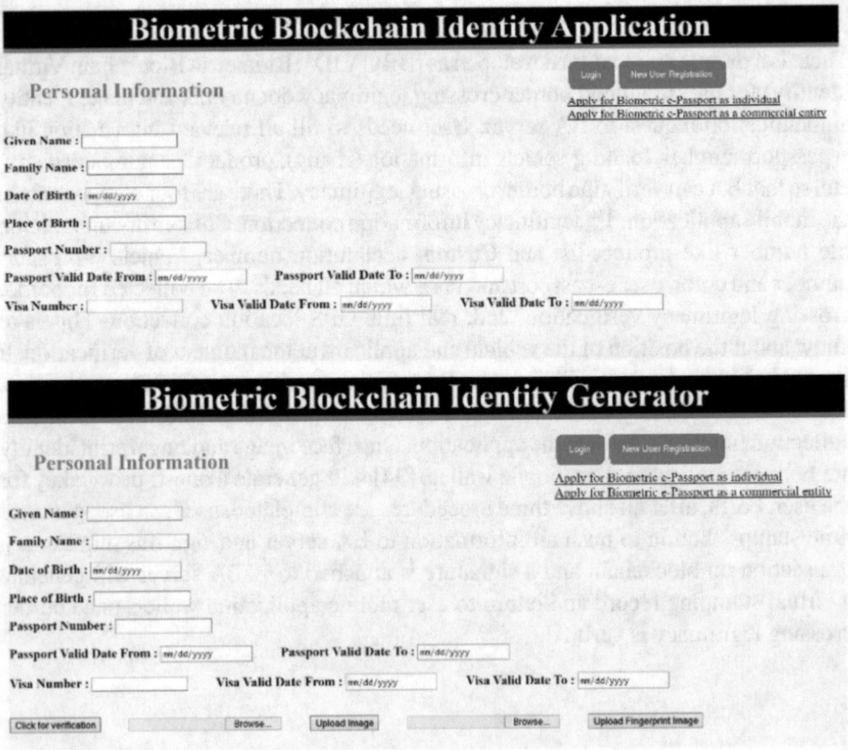

Fig. 3 Demonstration of e-passport application and generation website

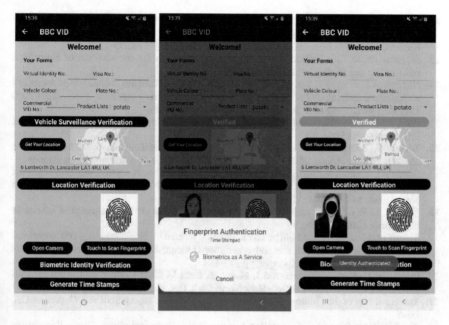

Fig. 4 Demonstration of mobile application for e-passport holder to push border crossing request and biometric identity authentication

References

1. Duggan C (2020) Lorry drivers could be slapped with £300 fines when travelling to Europe and back unless they have a Kent travel pass after Brexit. KentOnline. Accessed: 10 Sep 2021. [Online]. Available https://www.kentonline.co.uk/kent/news/plan-for-300-fines-to-stop-7-000-lorry-queue-235697/
2. Arpaia LAaP (2020) Experimental test of ECDSA digital signature robustness from timing-cattice attack. IEEE Instrument Meas Soc
3. Nakamoto S (2008) Bitcoin: a peer-to-peer electronic cash system. https://Bitcoin.org/en/Bitcoin-paper
4. Tijan E, Aksentijević S, Ivanić K, Jardas M (2019) Blockchain technology implementation in logistics. Sustainability 11(4)
5. Hackius NaP, Moritz (2017) Blockchain in logistics and supply chian trick or treat. Proceedings of the hamburg international conference of logistics, 23:3–18, ISBN 978–3–7450–4328–0
6. Xiang X, Wang M, Fan W (2020) A permissioned blockchain-based identity management and user authentication scheme for e-health systems. IEEE Access 8:171771–171783
7. Qadri YA, Nauman A, Zikria YB, Vasilakos AV, Kim SW (2020) The future of health-care internet of things: a survey of emerging technologies. IEEE Commun Surveys Tutorials 22(2):1121–1167
8. Fidas C, Belk M, Portugal D, Pitsillides A (2021) Privacy-preserving biometric-driven data for student identity management: challenges and approaches. Proceedings adjunct proceedings of the 29th ACM conference on user modeling, adaptation and personalization, 20
9. Wang R, Lin Z, Luo H (2018) Blockchain, bank credit and SME financing. Qual Quant 53(3):1127–1140
10. Craft DH (1983) Resource management in a decentralized system. ACM, Computer Laboratory, University of Cambridge

11. 'Bitcoindeveloper'. Transactions. Accessed: 5 Jun 2020. [Online]. Available: https://developer. bitcoin.org/devguide/transactions.html
12. Kleinrock L (1985) Distributed systems. Commun ACM 28(11):1200–1213
13. Stoianov ACaA (2017) Biometric encryption chapter from the encyclopedia of biometrics. Information and privacy commissioner, Ontario, Canada
14. Naganuma K, Suzuki T, Yoshino M, Takahashi K, Kaga Y, Kunihiro N (2020) New secret key management technology for blockchains from biometrics fuzzy signature. Proceedings 2020 15th Asia joint conference on information security (AsiaJCIS)
15. Patel VM, Ratha NK, Chellappa R (2015) Cancellable biometrics: a review. IEEE Signal Process Mag 32(5):54–65
16. Toutara F, Spathoulas G (200) A distributed biometric authentication scheme based on blockchain. Proceedings 2020 IEEE international conference on blockchain (Blockchain)
17. Sawant G, Bharadi V (2020) Permission blockchain based smart contract utilizing biometric authentication as a service: a future trend. Proceedings 2020 international conference on convergence to digital World—Quo Vadis (ICCDW)
18. Yang X, Li W (2020) A zero-knowledge-proof-based digital identity management scheme in blockchain. Comput Secur, 99
19. Kaga Y et al (2017) A secure and practical signature scheme for blockchain based on biometrics. In Information security practice and experience (Lecture Notes in Computer Science), vol 10701. Cham, Switzerland, Springer, pp 877–891
20. Lundkvist C, Heck R, Torstensson J, Mitton Z, Sena M (2016) UPort: a platform for self-sovereign identity. The BlockchainHub: North York, ON, Canada
21. Travel Identity of the Future—White Paper (2016) Technical report. ShoCard, Cupertino, CA, USA
22. The Inevitable Rise of Self-Sovereign Identity (2018) Technical report. Sovrin Foundation, Provo, UT, USA
23. Zwitter AJ, Gstrein OJ, Yap E (2020) Digital identity and the blockchain: universal identity management and the concept of the "self-sovereign" individual. Frontiers in Blockchain, 3
24. Al-Bassam M (2017) 'SCPKI: a smart contract-based pki and identity system'. Proceedings of the ACM workshop on blockchain, Cryptocurrencies and Contracts
25. W3C. Verifiable Credentials Data Model 1.0. 19 November 2019. Accessed: 10 Sep 2020. [Online]. Available https://www.w3.org/TR/vc-data-model/
26. Sporny M et al (2021) Decentralized identifiers (DIDs) v1.0 core architecture, data model, and representations. W3C. Aug 3rd, 2021. Accessed 12 Jan 2020. [Online]. Available: https:// www.w3.org/TR/did-core/
27. W3C. Use Cases and Requirements for Decentralized Identifiers. 17 March 2021. Accessed: 10 Sep 2020. [Online]. Available https://www.w3.org/TR/did-use-cases/
28. Tomko GJ, Soutar C, Schmidt GJ (1994) Fingerprint controlled public key cryptographic system. U.S. Patent 5541994, July 30, 1996 (Filing date: Sept. 7, 1994)
29. Juels A, Wattenberg M (1999) A fuzzy commitment scheme. In: Tsudik G (ed) Sixth ACM conference on computer and communications security. ACM Press, New York, pp 28–36
30. Juels A, Sudan M (2002) A fuzzy vault scheme. In: Lapidoth A, Teletar E (eds) Proceedings of IEEE international symposium on information theory, IEEE, Lausanne, 408
31. Chang D, Garg S, Hasan M, Mishra S (2020) Cancelable multi-biometric approach using fuzzy extractor and novel bit-wise encryption. IEEE Trans Inf Forensics Secur 15:3152–3167
32. Takahashi K, Matsuda T, Murakami T, Hanaoka G, Nishigaki M (2015) A signature scheme with a fuzzy private key. In: Malkin T, Kolesnikov V, Lewko A, Polychronakis M (eds) ACNS 2015. LNCS, Springer, Heidelberg, vol 9092, pp 105–126. https://doi.org/10.1007/978-3-319-28166-76
33. Sporny M et al (2020). JSON-LD 1.1A JSON-based serialization for linked data. W3C. July 16th 2020. Accessed 25 Jan 2020. [Online]. Available: https://www.w3.org/TR/json-ld/
34. 'Bitcoindeveloper'. Wallets. Accessed: 8 Jun 2020. [Online]. Available: https://developer.bit coin.org/devguide/wallets.html

Printed in the United States
by Baker & Taylor Publisher Services